The Macmillan Encyclopedia of Weather

The Macmillan
Encyclopedia of Weather

by Paul Stein

Macmillan Reference USA
An imprint of the Gale Group
New York • Detroit • San Francisco • Boston • Woodbridge, CT

Copyright © 2001 by Rosen Book Works, Inc.

First Edition

All rights reserved, including the right of reproduction in whole
or in part in any form.

All rights to this publication will be vigorously defended.

Macmillan Reference USA   Rosen Book Works, Inc
1633 Broadway   29 East Twenty-First Street
New York, NY 10019   New York, NY 10010

Library of Congress Cataloging-in-Publication Data

Stein, Paul, 1968-
    Macmillan encyclopedia of weather / by Paul Stein.
        p. cm.
    Includes bibliographical references and index.
    ISBN 0-02-865473-0 (hc. : alk. paper)
1. Meteorology--Encyclopedias, Juvenile. [1. Weather--Encyclopedias. 2. Meteorology--Encyclopedias.] I. Title.

QC854 .S74 2000
551.5'03--dc21

Cover photos: Tornado, D. Burgess Photographer © National Oceanic and Atmospheric Administration/Department of Commerce; Hurricane, Peter Dodge Photographer © AOML Hurricane Research Division; Snowstorm © National Oceanic and Atmospheric Administration/Department of Commerce.

Table of Contents

	Introduction	6
A	Acid Rain	9
	Adiabatic Process	10
	Advection	12
	Air	14
	Air Mass	17
	Albedo	20
	Alberta Clipper	22
	Anemometer	23
	Aristotle	24
	Atmosphere	26
	Aurora	28
B	Barometer	30
	Blocking Pattern	31
	Butterfly Effect	33
C	Catalina Eddy	34
	Climate	36
	Climate Change	39
	Cloud Classification	41
	Cloud Formation	42
	Cold Front	44
	Cold Wave	46
	Convergence	47
	Coriolis Effect	48
	Cyclogenesis	50
D	Degree Day	53
	Derecho	53
	Desertification	55
	Dew	57
	Dew Point	58
	Divergence	59
	Doppler Radar	60
	Downburst	63
	Drought	64
	Dryline	66
	Dust Storm	67
E	Easterly Wave	70
	El Nino	72
	Equinox	74
	Extratropical Cyclone	75
F	Flood	79
	Fog	84
	Franklin, Benjamin	86
	Freezing Rain	87
	Frost	88
	Frost Bite	90
	Fujita, Theodore	91
G	General Circulation	93
	Global Warming	94
	Graupel	97
	Greenhouse Effect	99
H	Hail	100
	Halo	102
	Haze	103
	Heat Exhaustion and Heat Stroke	104
	Heat Index	105
	Heat Transfer	106
	Heat Wave	108
	High Pressure	109
	Hurricanes	110
	Hydrological Cycle	118
	Hydrostatic Equilibrium	119
	Hygrometer	120
	Hypothermia	121
I	Ice Age	121
	Intertropical Convergence Zone	123
	Isopleth	125
J	Jet Stream	127
L	Lake Effect Snow	128
	La Nina	130
	Lightning	132
	Local Winds	135
	Low Level Jet	137
	Low Pressure	138
M	Meridional Flow	140
	Mesocyclone	140
	Mesoscale Convective Complex	142
	Meteorology	144
	Microclimate	148
	Mirage	150
	Monsoon	152
N	National Weather Service	155
	Nor'easter	157
	Numerical Weather Prediction	159
O	Occluded Front	161
	Oceans	163
	Orographic Lift	165
	Overrunning	166
	Ozone	168
P	Pollution	170
	Precipitation	172
R	Radar	174
	Radiation	176
	Rain	177
	Rainbow	179
	Rain Shadow	180
	Rossby, Carl-Gustav	182
S	Satellite	183
	Scales of Motion	185
	Scattering	187
	Seasons	189
	Sleet	190
	Snow	192
	Solstice	193
	Sounding	194
	Stability	196
	Stationary Front	198
	Storm Chasing	199
	Storm Surge	201
T	Temperature	203
	Thermometer	205
	Thunderstorms	206
	Tornadoes	216
	Trade Winds	225
	Turbulence	227
U	Upper-Level Ridge	229
	Upper-Level Trough	230
	Urban Heat Island	232
V	Visibility	233
W	Warm Front	235
	Waterspout	236
	Water Vapor	238
	Weather	241
	Weather Extremes	243
	Weather Forecasting	247
	Weather Map	250
	Weather Modification	250
	Wind	252
	Wind Chill	254
	Wind Profiler	255
	Wind Shear	256
	Winter Storms	258
Z	Zonal Flow	265

INTRODUCTION

On May 2nd, 1999, forecasters became increasingly concerned as they reviewed computer-generated maps showing predicted weather conditions over the central United States for the following afternoon. The forecasts they examined were created by atmospheric models—extraordinarily complex computer programs that simulate the movement of weather systems for days into the future. On the afternoon of May 2nd, the computer models were showing a frightening potential for violent weather in the southern Plains for the next day.

As meteorologists continued to monitor computer model output into the morning of May 3rd, they also tracked cloud patterns from satellite imagery. The satellite photos clearly showed a strong weather system steadily advancing towards the central United States. By that afternoon, large thunderstorms began erupting over central Oklahoma. Forecasters used Doppler radar to monitor wind patterns inside the thunderstorms as they rapidly intensified.

As tornadoes began to strike Oklahoma on May 3rd, television and radio in Oklahoma City interrupted programming and devoted live coverage to the developing situation. Television stations were able to track the tornado-producing thunderstorms using their own Doppler radars, and by sending out storm spotters in vehicles and helicopters. One tornado—a monster—quickly became the focus of attention as it roared towards Oklahoma City. Unprecedented live media coverage of the event allowed the public to take shelter well in advance as the mile-wide tornado emerged from open farmland and into the subdivisions in the southern part of the city.

Over thirty people were killed and hundreds more were injured as the tornado damaged or destroyed over 8,000 homes and businesses across southern Oklahoma City. The death toll, however, would have been much higher were it not for the efficiency of electronic communication and the dedication and skill of weather forecasters, both on and off the airwaves.

Entering the twenty-first century, the science of meteorology is undergoing a revolution—and the historic weather event mentioned above serves to indicate just how meteorology has fundamentally changed through the last few decades.

Introduction

ON THE FRONTIER OF TECHNOLOGY
Computers, satellites, radar, and electronic communication have played major roles in the transformation of atmospheric science, while also hinting at how meteorology may continue to evolve.

As we head into the twenty-first century, computerized weather prediction is becoming more and more accurate and detailed. Already, some computerized weather predictions are employing artificial intelligence, allowing the computers to learn from their forecasting mistakes and grow more skilled over time. While subject to important limitations and certainly not infallible, atmospheric models promise to relieve the task of day-to-day temperature and weather forecasts that have traditionally been the staple of meteorologists for the last one hundred years. This will free them to concentrate on issuing life-saving severe weather warnings, and resolving important small-scale details in local weather patterns that still evade the computers' predictive abilities.

Weather satellites, first launched in the 1960s, have become indispensable in the forecasting process. They take photographs from space that allow meteorologists to see the large-scale weather systems moving across oceans and continents. Satellites are already evolving into not only a tool for taking photographs of clouds from high above, but also as an instrument for remotely sensing the temperature and moisture content at many levels of the atmosphere. This critical information allows meteorologists to track weather systems high in the atmosphere, not just at ground level. For most of the twentieth century, this data has only been available through the daily launch of instrument packages carried aloft by balloons. Satellites offer hope for a much more efficient, cheaper and more detailed source for this vital information. Data so gained will be fed back into computer models, resulting in further increases in forecasting accuracy and detail.

In the late twentieth century, doppler radar has been the third technological revolution in meteorology. While giving meteorologists extremely detailed images of the location and intensity of rain and snow, doppler radar has been most important in allowing meteorologists to see the swirling winds inside thunderstorms that lead to tornadoes. Indeed on the afternoon of May 3rd, meteorologists began issuing life-saving tornado warnings, in large part based on the valuable information obtained as doppler radars scanned developing thunderstorms.

UNDERSTANDING LONG-TERM WEATHER CHANGES
The same technology that is rapidly becoming more accurate in weather fore-

casting and tracking also assists scientists in understanding the long-term changes in weather and climate around the planet. These changes, less immediately spectacular than a raging hurricane or tornado, are even more threatening as they progress almost imperceptibly over time. The years 1998 and 1999, the warmest two years of the twentieth century in the United States, strongly hint that the earth's climate is changing in profound ways. It is on this front, the understanding of global climate, that the revolution in meteorology will confront its greatest challenge in the years ahead.

ABOUT THIS BOOK

This book will help you, the student, understand the changing face of meteorology—both the natural phenomena of Earth and the ways we have devised to study them. From gaining a more thorough understanding of the science behind television weather reports, to learning about the processes that govern the greenhouse effect and global warming, the study of meteorology draws upon and offers insight into a variety of human experience. Interwoven with weather are strands from chemistry, physics, mathematics, computer science, engineering, oceanography, biology, and media. Weather, in turn, affects everything from aviation and travel to health, the economy, sporting events and other outdoor activities, as well as the balance of plant and animal life and the fate of civilizations.

Learning about weather can transform the sky into a daily source of wonder. It also allows you to better understand, and even play an informed role in, the growing debate about the human influence on climate. Weather can alter life. To understand weather is to understand the changing earth, our relation to it, and the role we play in determining our own future.

ACID RAIN

Acid rain forms when industrial pollutants are released into the air and mix with cloud and rain droplets to produce acids. Acid rain can damage trees, wildlife, and buildings, and occurs most often downwind of power plants and industrial areas.

The pollutants mainly responsible for acid rain are sulfur dioxide and nitrogen oxides. These chemicals are produced from the burning of fossil fuels such as oil, natural gas, and coal. Automobile emissions are one significant source of pollutants that contribute to acid rain. As the wind spreads the airborne chemicals, often for many hundreds of miles from their origins, they react with water in cloud droplets, rain, and snow to produce sulfuric and nitric acid. Polluted clouds can drift over mountains, exposing high-altitude forests to acidic moisture. When the acidic moisture falls back to the ground in rain, it is absorbed by the soil or runs off into lakes, rivers, and other bodies of water. Industrial pollutants can also fall to the ground without having mixed with water, in a process known as dry deposition. Rain or melting snow on the ground reacts with the deposited pollutants to form the same acids found in acid rain.

Acids are sour-tasting and often corrosive. The acidity of a substance is measured on the pH scale. It ranges from 0 to 14; acids have a pH below 7, and alkalines (bases) have a pH value above 7. A neutral substance has a pH of exactly 7. The scale is exponential, so a liquid with a pH of 4 is ten times more acidic than one with a pH of 5, and 100 times more acidic than one with a pH of 6. Most precipitation is naturally slightly acidic, with a pH of 5.6. Thus, precipitation must have a pH value below 5.6 to be considered acid rain. Some rainfall over the northeastern part of the United States has been measured with a pH as low as 4.0.

Factories and power plants can emit pollutants into the air, causing rain to become acidic.

When the sulfuric and nitric acid in rainwater flows into lakes and streams, it can injure or kill aquatic plant and animal life. When smaller lakes are exposed to acid rain over a long period of time, they may become completely unable to support life. Limestone in soil can neutralize the acidity, but forests growing in limestone-deficient areas are vulnerable to damage from years of absorbing acid rain runoff. Studies have shown that acid rain and other airborne pollutants have damaged at least half of the trees in Germany's 2000-square-mile Black Forest. Major tree damage has also been observed in Québec, Canada, due in large part to polluted air originating in the United States. Over years, the acids in rainfall can also corrode buildings and outdoor statues.

In 1990 the United States Congress passed legislation called the Clean Air Act aimed at reducing the industrial emissions that cause acid rain. In support of this effort, the National Atmospheric Deposition Program continually collects data on acidic deposition at over 200 sites around the country. *See also: Air; Rain; Snow; Wind.*

Adiabatic Process

The adiabatic process is the method by which air parcels rise and fall in the atmosphere without exchanging energy with the outside air. The adiabatic process helps explain how rising air forms clouds and how sinking air becomes warmer. When describing the movement of air, meteorologists often use the concept of an air parcel. An air parcel is just an imaginary bubble, or volume, of air. In an adiabatic process, air outside the parcel does not mix with air inside the parcel.

One of the most important aspects of the adiabatic process is that it gives meteorologists a way of predicting how the temperature of rising or falling air will change. When an air parcel rises, there are fewer air molecules to weigh down on the air parcel from above. Therefore, a rising air parcel moves into lower atmospheric pressure. Because the atmospheric pressure around the air parcel is lower, the air parcel expands outward. When the parcel expands, the air molecules inside the parcel become farther and farther apart. The molecules slow down, and the temperature of the parcel drops.

The rate at which a rising air parcel's temperature cools depends on the amount of water vapor in the air parcel. In air, water vapor is continually changing from gaseous water to tiny liquid water droplets and back again. The process whereby water vapor changes into liquid water is called condensation, and this process releases energy. The process whereby liquid water changes into water vapor is called evaporation, and this process requires energy.

When a parcel of air first begins to rise adiabatically from the ground into the atmosphere, the temperature of the parcel will still be relatively warm. Water vapor molecules in the air will move around rapidly and will not easily adhere to one another, or condense, when they collide. Therefore, the rate of evaporation in the air parcel will exceed the rate of condensation. Since evaporation removes energy from the air, the air parcel will cool at a relatively faster rate as it rises. This rate is called the dry adiabatic rate and is equal to 5.5°F per 1000 feet or 10°C per 1000 meters.

The adiabatic process helps explain how clouds form in rising air.

As an air parcel rises higher and higher into the atmosphere, it continues to cool and expand. Once the air parcel cools far enough, the rate of condensation inside the air parcel will exceed the rate of evaporation, and liquid water droplets will begin to form a cloud. Since condensation adds energy to the air, the rising air parcel will cool more slowly once a cloud forms. This slower rate is called the moist adiabatic rate and averages around 3.3°F per 1000 feet or 6°C per 1000 meters.

A sinking air parcel, by contrast, compresses as it falls into higher and higher atmospheric pressure. The density of air molecules increases as the air parcel approaches the ground, causing the air parcel to constrict. As the volume of the sinking air parcel decreases, the molecules inside move around faster and faster. The temperature of the parcel therefore rises. Unlike rising air, there is only one rate at which the sinking air's temperature changes. Sinking air always warms at the dry adiabatic lapse rate. Because molecules inside the air parcel move faster when air sinks, evaporation exceeds condensation.

Because of the adiabatic warming of sinking air, wind that blows down from mountains or other significant elevations is often warming. As an example, in the large cities of the northeastern United States the hottest summer days occur when the wind is blowing from the west or northwest. This is because air blowing from this direction descends the Appalachian Mountains. Air blowing from the southwest flows parallel to the mountains, while air blowing from the south comes from the relatively cooler ocean water.

Cities along the West Coast experience similar, and at times more dramatic warming. When winds blow from the east, they descend thousands of feet from mountain chains that lie just inland. Air that blows in from this direction can warm many tens of degrees. If accompanied by strong, gusty winds, these adiabatic warming events are known as Santa Anas. These are dangerous conditions, since fires can spread rapidly in the strong, dry, and hot flow of air. ***See also: Air Mass; Atmosphere; Cloud Formation; Oceans; Temperature; Water Vapor; Wind.***

ADVECTION

As wind blows horizontally over the earth's surface, it often brings changes in atmospheric properties, such as temperature or moisture, in a process known as advection. Most people, for example, have experienced cold advection in the fall or winter as a cold front crosses their location. Gusty, northerly winds transport, or advect, colder air into their area after the cold front passes. Meteorologists look for areas of temperature or moisture advection to aid in forecasting weather over a given location.

The magnitude of advection depends on a couple of factors. One is wind speed. Stronger winds will advect a larger quantity of air than weaker winds, transporting more of an atmospheric property over a given period of time. Another factor that affects the magnitude of advection is the difference in the air temperature or moisture from one location to another. The larger these differences are, the greater the magnitude of advection. Also important is the direction of the wind in relation to the difference in air temperature or moisture. For example, if the air becomes colder from south to north, then a wind from the west will blow parallel to the temperature difference and no advection will occur. However, if the wind blows from the north, it will advect much colder air southward.

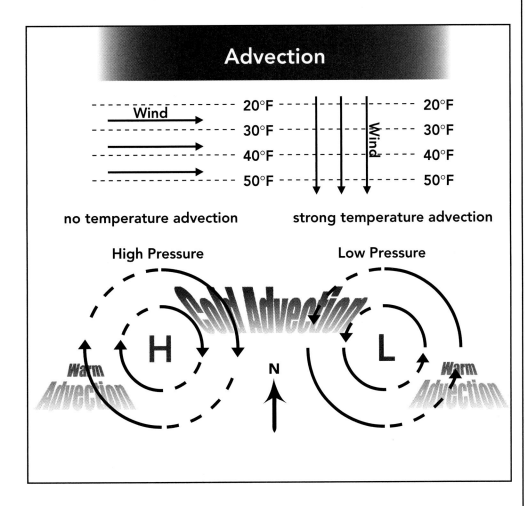

Temperature advection depends on the wind direction. Areas of cold and warm advection often occur around high and low pressure areas.

Areas of temperature and moisture advection occur in conjunction with large-scale weather systems. Air circulates in a counterclockwise direction around areas of low pressure, called cyclones by meteorologists. West of the cyclone, northerly winds advect drier, colder air southward. East of the cyclone, southerly winds advect warmer, moister air northward. Advection also occurs around areas of high pressure, called anticyclones. Around anticyclones, air circulates in a clockwise direction. Therefore, on the west side of high pressure, winds advect warmer, more moist air northward. On their east side, winds advect colder, drier air southward. Areas of high and low pressure are often situated adjacent to one another, so each system supports the same areas of temperature and moisture advection.

Advection can occur both at the earth's surface and at high levels of the atmosphere. Areas of warm and cold air advection near the earth's surface are measured to predict changes in temperature. The larger the advection of surface cold air, for instance, the lower the temperature might fall over a given area. At high levels of the atmosphere, temperature advection can change the stability of the atmosphere. An unstable atmosphere is one in which air easily rises, forming clouds and precipitation, while in a stable atmosphere, air does not easily rise. Air becomes unstable when the temperature decreases rapidly from the earth's surface upward to high levels of the atmosphere. When winds advect colder air aloft, this can increase the vertical temperature difference and cause the atmosphere to become unstable. At either low or high levels of the atmosphere, the advection of moisture into an area can help meteorologists pinpoint where areas of rain or snow might develop. *See also: Air; Cloud Formation; Cold Front; High Pressure; Low Pressure; Temperature; Wind.*

Air

Air consists of a mixture of gases, microscopic particles, pollutants, and dust, the concentrations of which vary relative to altitude, season, and location. While nitrogen and oxygen together account for over 99 percent of the atmosphere's volume, other important constituents of air include water vapor, carbon dioxide, and ozone.

The composition of the air we breathe today is very different from the air that existed billions of years ago. As the earth was forming, the atmosphere consisted mainly of hydrogen, helium, and related compounds-the most abundant gases in the universe. These gases, however, are very light and probably escaped into space over time.

Large amounts of gas, however, remained trapped in the solid earth after it formed. These gases, including water vapor, carbon dioxide, and nitrogen, escaped from the earth's hot interior through volcanoes, vents, and cracks in the ground. Over hundreds of millions of years, this outgassing resulted in the earth's first atmosphere. Increasing amounts of water vapor gave rise to clouds and precipitation, which in turn led to the formation of lakes and oceans.

These expanding bodies of water absorbed increasing amounts of the carbon dioxide, leaving nitrogen as the largest constituent of air. The reaction of ultraviolet energy with water vapor in a process called photodissociation created oxygen, which in turn supported the beginnings of plant life. As plant life developed over millions of years, photosynthesis created even more oxygen. The air we breathe today has reached a relative state of equilibrium. The concentrations of a few of the more abundant gases in air are regulated by natural organic processes—decay of animal and plant life, breathing (respiration), and photosynthesis.

Over the last few decades, scientists have focused attention on some of the more scant ingredients of air. Carbon dioxide, by volume a minor component of air, plays an important role in the regulation of global temperature. Carbon dioxide is a greenhouse gas, meaning that it is an efficient absorber of outgoing energy from the earth. The burning of fossil fuels, including oil, coal, and natural gas, has caused a measurable increase in the amount of carbon dioxide in the atmosphere over the last fifty years. This increase may be contributing to a warming of the atmosphere. How this global warming will affect weather patterns is the subject of ongoing research.

Ozone molecules, accounting for a tiny fraction of the volume of air, play different roles depending on their location. Near the ground, ozone is an unhealthy

The air we breathe consists of different gases, microscopic particles, pollutants and dust.

pollutant that collects over cities under stagnant conditions. In the higher reaches of the stratosphere, however, ozone absorbs harmful ultraviolet radiation from the sun. Molecules called chlorofluorocarbons (CFCs), introduced into air through the manufacture and use of certain consumer products, are likely contributing to a thinning of the earth's protective stratospheric ozone.

Air becomes unhealthy when it contains too many pollutants.

Water vapor is an extremely important gas found in air. The process whereby water vapor changes from a gaseous state to a liquid state is known as condensation. This process forms clouds that produce rain and snow, and results in an exchange of heat between water vapor and air. This heat transfer is an important factor in the formation of storms. Since water vapor is an efficient absorber of outgoing radiation from the earth, it too is a significant greenhouse gas. Therefore, like carbon dioxide, it helps regulate the temperature of the atmosphere.

All these molecules—nitrogen, oxygen, water, carbon dioxide—together with dust, pollen, pollutants, and other gases and particles, are the chemical ingredients of air. The collective behavior of these air molecules, according to fundamental physical and thermal laws, results in weather.

The temperature of air is just a measure of molecular kinetic energy, the average rate that the constituent molecules move. Air pressure indicates the force per unit area exerted by these moving molecules. Wind is the collective movement of millions of these molecules.

A simple equation called the gas law states that air pressure is proportional to temperature and density. In the atmosphere, areas of high pressure can be either warm or cold, depending on the density of the air. In winter, frigid high-pressure systems moving out of polar regions contain a high density of air molecules. In summer, on the other hand, warmer high-pressure systems moving across the United States contain a lower density of air molecules. Air density, therefore, is the key to understanding the relationship between air pressure and air temperature. *See also: Atmosphere, Cloud Formation, Global Warming, Heat Transfer, High Pressure, Ozone, Precipitation, Radiation, Temperature, Water Vapor.*

AIR MASS

An air mass is a large expanse of air with a relatively uniform density of air molecules. It is usually accompanied by a characteristic type of weather, known as air mass weather, which depends on the temperature and moisture of the air mass. Meteorologists predict weather changes by tracking air masses as they spread from the area over which they formed, called their source region.

FORMATION OF AIR MASSES

Air masses form in regions of relatively stagnant air, where winds are calm over a large and relatively flat expanse of land and water. Air masses frequently originate from high-pressure areas since these weather systems often move slowly and contain light winds. As air spends more and more time over the same region, it acquires the temperature and moisture characteristics of the ground or water underneath it. Air that collects over tropical ocean water, for instance, becomes warm and humid. As the air takes on a relatively uniform character over a large area, an air mass is born.

Subtropical areas (around 30° north and south latitude) and polar regions are the two primary air mass source regions. Over these regions, the air flow at high altitudes of the atmosphere tends to converge, or come together, and increase the amount of air aloft. The more air that gathers overhead, the higher the air pressure becomes at the earth's surface.

Air masses extend over hundreds or even thousands of miles and can be of varying thickness depending on the density of the air. When air has a high pressure and low temperature, for example, air molecules will move much more slowly and will compress together. This is why cold, high-pressure areas will contain relatively dense air. The dense air settles near the earth's surface and the thickness of such an air mass may be only hundreds of feet. In contrast, a warmer air mass, with air molecules that move faster and spread farther apart, may extend thousands of feet in height.

CLASSIFICATION OF AIR MASSES

Meteorologists label air masses using a two-letter code. The first letter, traditionally written in lower case, identifies the source region of an air mass. The letter **c** is used for an air mass that originates over a continental land area, while the letter **m** refers to an air mass that forms over a maritime, or ocean, environment. The second letter of the two-letter code is written in upper case and classifies the air mass by the latitude of its source region. The letter **P** indicates a polar air mass, while a **T** denotes a tropical air mass. Used in combination, these letters provide

Different kinds of air masses travel into the United States from their source regions.

meteorologists with a shorthand method of categorizing air masses.

A cP, or continental polar air mass, most frequently develops in the high latitudes of the Northern Hemisphere in winter. A cP air mass accompanies large high-pressure areas that form over places such as Siberia, Alaska, and northwestern Canada. Long nights and snow-covered ground allow air to grow very cold. As the cP air mass migrates southward, it brings cold, dense air into southern Canada and the continental states. Outbreaks of cP air result in wintertime cold waves with record low temperatures occurring in the coldest cP air masses. Since these air masses are so shallow, the Rocky Mountains act as a barrier over which cP air cannot usually spread. Occasionally, however, a large area of high pressure and its accompanying cP air spreading south from Canada take a more westward track. When this occurs, frigid conditions and snow can spread into normally milder sections of the Pacific Northwest.

An air mass that originates over the ocean in northern latitudes is known as mP air. This kind of air often forms over the North Pacific and spreads into the West Coast of the United States in winter. Owing to its oceanic origins, it contains more moisture than an air mass originating over land. While it can be cool, the relatively milder ocean water prevents mP air from becoming as cold as air that gathers over land at similar latitudes. An mP air mass causes showers and mountain snow as it spreads into the Pacific states. Far less frequently, mP air over the North Atlantic can flow south and west into the northeastern United States. Residents of this area are familiar with the term "back door cold front," referring to the opposite direction from which most fronts approach. These fronts mark the leading edge of mP air invading the northeast and its attendant cloudy, chilly, and damp weather.

Air that forms over the tropical oceans is labeled mT air. It is usually warm and humid, often accompanying heat waves in the eastern part of the United States in summer. Owing to the large quantities of tropical moisture in an mT air mass, it often brings scattered thunderstorms in the afternoon as the sun heats the ground. The West Coast also experiences mT air. Occasionally winds direct warm, moist air northward from the tropical Pacific in wintertime. California suffers some of its worst flooding when mT air spreads in from the Pacific. In the summer, mT air can move northward from the Mexican coast into the desert Southwest. These normally dry areas experience afternoon thunderstorms as the warm, moist tropical air is heated by the sunshine.

Finally, cT air forms over northern Mexico. Since this part of North America is relatively narrow, cT air masses typically do not gain the breadth of air masses that form over large tracts of land or ocean. When cT air spreads northward, hot and dry conditions are experienced across the central and southwestern United States. *See also: Air; Cold Front; Cold Wave; Heat Wave; High Pressure; Oceans; Snow; Thunderstorms.*

Albedo

The albedo of an object is the percentage of electromagnetic energy that the object reflects as compared to the amount that strikes it. Electromagnetic energy is known by scientists as radiation energy, or simply radiation. Meteorologists use the concept of albedo in relation to the sun's radiation energy as it strikes the earth. The albedo of the ground affects how warm the air becomes directly above it.

Snow has a high albedo since it reflects most of the sunlight that hits it.

Bright or white surfaces tend to reflect radiation, and therefore have higher albedos than darker surfaces, which absorb radiation. Since radiation is reflected away from bright surfaces, the temperature of these surfaces tends to be lower as well. Air that comes into contact with a substance that has a high albedo will therefore not warm as much as air in contact with a substance with a low albedo. Other factors, however, are also important in regulating air temperature at a given location: how far north or south the location is, how much moisture is in the air, and local terrain, among other things. Therefore, air is not always warmer over a darker ground surface and cooler over a brighter one.

The effect of albedo can be seen and felt with fresh snow cover. In winter the sun is low in the sky and its rays strike the earth at a much lower angle. Despite the low sun angle, skiers can easily receive a sunburn in the middle of winter. This is due in part to the high albedo of the snow; its white color reflects much of the solar energy that strikes it. (Multisided snow crystals provide numerous surfaces from which sunlight can be reflected. Ice has a much lower albedo than snow, owing to its generally darker color and flat surface.) Some of this reflected energy is absorbed into the skin, which causes sunburn. The effect of albedo is also seen in hot climates. In desert and tropical regions, houses, buildings, and clothing are often white or bright in order to limit the amount of heat that is absorbed. *See also: Air; Radiation; Snow; Temperature.*

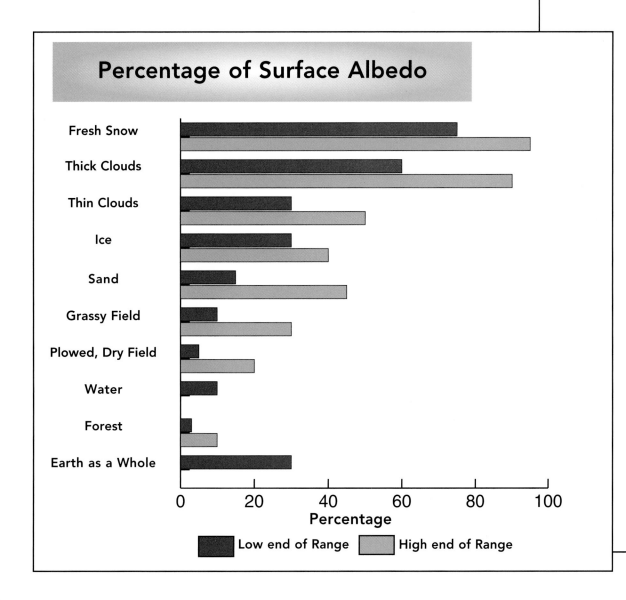

The table at the left shows the albedo of various surfaces. Different surfaces have a wide range in albedo.

Alberta Clipper

The term "Alberta clipper" refers to a fast-moving, small area of low pressure that moves through parts of the northern United States. The term refers to the location from where these storms originate: the province of Alberta in western Canada. Alberta clippers usually occur in winter and can drop several inches of snow along and to the north of their track.

Alberta clippers form when winds at high altitudes in the atmosphere blow from the northwest, out of the Canadian Rockies, and down into the midwestern United States. Storms that form in Alberta in this northwesterly flow follow the winds southeastward. A typical track for this kind of storm extends from western Canada into the upper Midwest, then down into the Ohio Valley or lower Great Lakes region before turning back to the northeast through New England. If the northwesterly flow persists for a week or more, several Alberta clippers may form and move along in succession, each one following the other at a rate of one every several days.

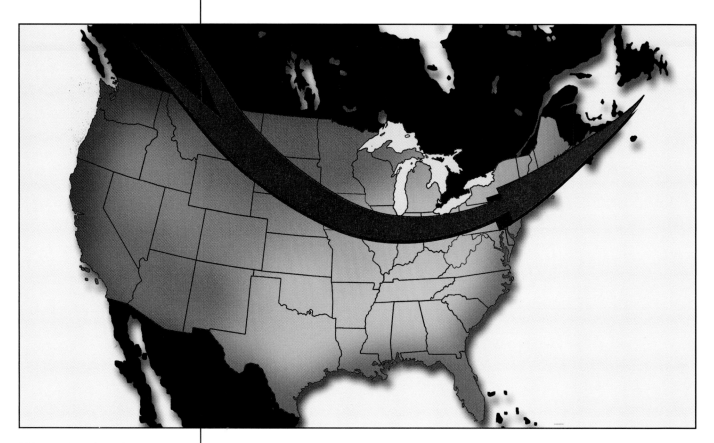

This map shows the path commonly taken by an Alberta Clipper as it moves across the northern United States.

Although these storms are small and fast-moving, they can cause several hours of heavy snow and gusty winds at any given location as they move along. Snow will occur in the cold air along and north of the path of the Alberta clipper. South of the track of the storm, southerly winds often result in mild, breezy conditions with rain showers.

Even though most Alberta clippers are considered no more than a nuisance, they can sometimes play a role in the formation of a much larger storm. This can happen if an Alberta clipper dropping southward from Canada combines with another storm moving northward from the southern United States. Under the right conditions, the two storms can combine and form a much larger storm along the East Coast. *See also: Air; Atmosphere; Low Pressure; Rain; Snow.*

ANEMOMETER

An anemometer is an instrument that measures wind speed. Most anemometers today consist of three or four metal cups attached to a vertical rod. The rod can be mounted on a rooftop or attached to a pole set in the ground. As the wind hits

The cups of an anemometer whirl around when the wind blows, recording the speed of the wind.

the metal cups, they rotate. The rate of rotation depends on how fast the wind is blowing. A system of gears or electronically linked sensing devices enables a meteorologist to read the wind speed from an anemometer without having to be outside.

Meteorologists use anemometers to measure wind speed for a variety of reasons. Wind is a phenomenon that affects anyone outdoors; high wind speeds can cause damage and can threaten lives. Airports often have a network of anemometers that measure wind speed and direction at various points along the runways. Planes need this wind information as they take off and land. Sailboaters also rely on anemometers to know how fast the wind is blowing. Some meteorologists who study atmospheric pollutants use the data from anemometers to understand how gases and airborne particles move through the atmosphere.

There are several different kinds of anemometers. An older variety is called a pressure plate anemometer. With this instrument, wind blows against a suspended plate. The angle to which the plate is pushed by the moving air corresponds to the strength of the wind. A pressure tube anemometer is one where wind is measured as air blows into or across an open tube. The pressure of the air inside the tube varies, depending on the strength and direction of the wind. An aerovane measures both wind speed and direction, and looks like a model airplane mounted on a shaft. Wind blows through propellers attached at the nose of the aerovane; the rate at which the propellers spin depends on the wind speed. Wind blowing against the tail section causes the aerovane to rotate on its shaft, indicating the direction from which the wind is blowing.

Buildings, trees, and other objects disrupt wind flow at ground level. Wind funneling through closely spaced buildings can blow at higher speeds, whereas wind blowing through a line of trees will slow. Therefore, all official wind speed readings are taken by anemometers at a height of around ten meters above the ground. *See also: Air; Atmosphere; Wind.*

ARISTOTLE

The first known book on meteorology was written by Aristotle (384-322 BCE), a Greek philosopher. The word "meteorology" actually originates from the book's title, *Meteorologica*, referring to the study of objects that fall from the sky. In this book, Aristotle speculates on the nature of weather, making both wildly erroneous and remarkably perceptive hypotheses.

Aristotle was a student of the famous philosopher Plato and was a tutor to Alexander the Great. He was best known for his extensive writings on nature, including observations on biology, zoology, and astronomy, as well as teachings on ethics, politics, and metaphysics. His writings on meteorology were the major influence on the science for almost 2,000 years, dispelled only by scientific and mathematical advances in Europe in the eighteenth century.

In *Meteorologica*, Aristotle attempts to explain wind as a component part of a living planet rather than the movement of air. He proposes that the earth, as a life entity, has "exhalations" similar to the breath of a human being. These dry and moist exhalations, he claims, flowed from vents and cracks in the earth's crust. Gathering together like streams merging into rivers, these exhalations combine to form the wind that blows across the earth. "It is absurd," Aristotle notes in *Meteorologica*, "that this air surrounding us should become wind when in motion."

Aristotle, a Greek philosopher, wrote the first known book on meteorology over 2,000 years ago.

As inaccurate as this was, Aristotle developed at least one theory that was ahead of its time. In *Meteorologica*, Aristotle describes how the sun's heating results in the movements of masses of air and moisture: "So we get a circular process that follows the movement of the sun . . . We must think of it as a river, flowing up and down in a circle and made up partly of air, partly of water." Now we know this phenomenon as the hydrological cycle: the circulation of water from lakes, rivers, and oceans, into the atmosphere through evaporation, then back to the land again as rain and snow.

A student of Aristotle, Theophrastus, also wrote about meteorology in his *Book of Signs*, written about 300 BCE. He claims certain natural signals, including clouds and insect behavior, can be used to foretell the weather. Like Aristotle's *Meteorologica*, Theophrastus' *Book of Signs* was used as a significant meteorological text for centuries thereafter. ***See also: Air; Hydrological Cycle; Meteorology; Oceans; Rain; Snow; Weather; Wind.***

ATMOSPHERE

Surrounding the earth's surface, our atmosphere is a layer of gases and microscopic particles that extends up for several hundred miles to an indefinite threshold with space. The earth's atmosphere consists primarily of two gases—nitrogen and oxygen—with lesser amounts of argon, ozone, and other gases, along with dust, water vapor, and suspended particles.

The atmosphere consists of different layers of air.

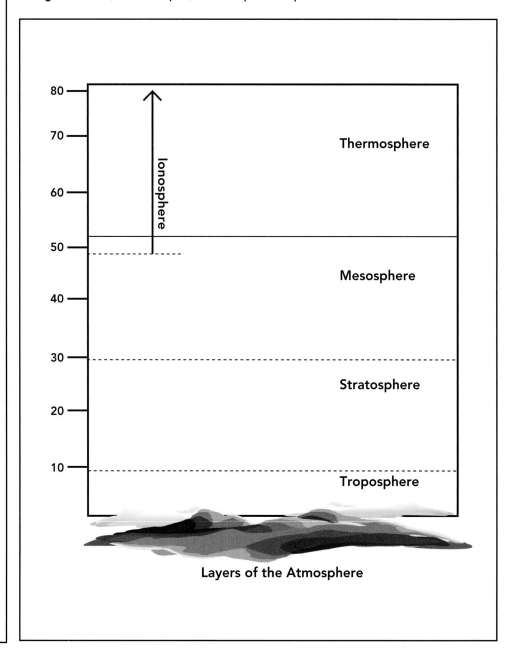

Layers of the Atmosphere

Layers of the Atmosphere

Meteorologists divide the atmosphere into horizontal layers, each characterized by different concentrations of gases and distributions of temperature. The layer closest to the ground is the troposphere. Most of the weather that we observe occurs in this layer. In the troposphere, air temperature and pressure generally decrease as height increases. The rate at which temperature decreases, known as the lapse rate, averages to around 3.6°F per 1000 feet. This value varies, depending on whether the air is dry or moist. At 35,000 feet, the height at which commercial jets often cruise, the air temperature can be as low as -50°F.

At an average height of around seven miles (36,960 feet), the temperature stops falling with increasing altitude. This is the start of the stratosphere, a layer of the atmosphere extending to around thirty miles above the earth's surface. Through most of this layer, the temperature increases with increasing height. Thus, any air that rises in the stratosphere quickly becomes colder and more dense than the surrounding atmosphere and sinks back to its original altitude. Since the stratosphere resists upward air currents, it is known as a stable layer of the atmosphere. The stratosphere is also home to the vast majority of the ozone in the atmosphere. Ozone, while behaving as a pollutant in the lower troposphere, plays a protective role in the stratosphere by absorbing harmful incoming ultraviolet radiation from the sun.

Above the stratosphere, the temperature begins to drop again. This indicates the next highest layer, the mesosphere, which extends above fifty miles in altitude. Very little air exists at this height. Nearly 99.9 percent of all the molecules in the atmosphere exist below the mesosphere. The air pressure here is less than 1/1000 of that at the earth's surface. At the top of the mesosphere, the temperature of this extremely thin air can be as low as -120°F.

The outermost atmospheric layer, the thermosphere, exists above the mesosphere. Once again the temperature rises with increasing height. At this layer, air molecules are very scarce. At the outer edge of the mesosphere, an air molecule must travel an average of 1,000 meters before colliding with another air molecule; this distance is only 0.00001 cm at the earth's surface. The diffuse edge of the atmosphere, where air molecules escape into space, is called the exosphere and is about 600 miles above the earth's surface.

Another layer, the ionosphere, spans the thermosphere and mesosphere. Through this layer, cosmic rays and ultraviolet radiation strip electrons from air molecules, so ions and free electrons exist in abundance. The ionosphere reflects shortwave and AM radio transmissions back to the earth, where people can receive these transmissions many hundreds or even thousands of miles from where they originated.

Evolution of the Atmosphere

The present composition of the atmosphere evolved from a much different state billions of years ago. As the earth was forming, the atmosphere consisted mainly of hydrogen, helium, and related compounds—the most abundant gases in the universe. These gases, however, are very light and probably escaped over time into space.

Large amounts of gas, however, remained trapped in the earth after it formed. These gases, including water vapor, carbon dioxide, and nitrogen, issued from the earth's hot interior through volcanoes, vents, and cracks in the ground. Over hundreds of millions of years, this outgassing resulted in the earth's first atmosphere.

Increasing amounts of water vapor gave rise to clouds and precipitation, which in turn led to the formation of lakes and oceans. These expanding bodies of water absorbed increasing amounts of carbon dioxide, leaving behind greater and greater amounts of nitrogen. Ultraviolet energy reacting with water vapor in a process called photodissociation created oxygen, which in turn supported the beginnings of plant life. As plant life developed over millions of years, photosynthesis created even more oxygen. This led to the atmosphere that exists today, consisting primarily of nitrogen and oxygen. *See also: Oceans; Ozone; Radiation; Temperature; Water Vapor.*

Aurora

The aurora is a luminous display of color, often shimmering or glowing in appearance, in the upper atmosphere. It is visible at night and occurs most often at high latitudes. In the Northern Hemisphere it is known as the aurora borealis; in the Southern Hemisphere it is called the aurora australis. It is caused when energized particles from the sun interact with atmospheric gases.

Interactions That Form the Aurora

The sun continuously emits a stream of high energy particles called the solar wind. Moving at a speed of 250 miles per second, the protons and electrons in the solar wind take about two days to reach the earth. The particles, also known as plasma, then interact with the earth's magnetic field.

The magnetic field of the earth is similar to that of a bar magnet. Magnetic field lines form a loop around the earth, exiting near the South Pole and entering

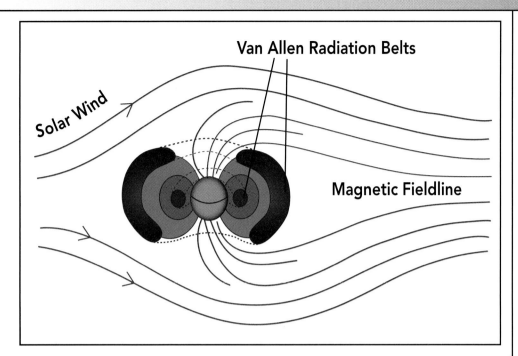

The Aurora forms when the solar wind interacts with the earth's magnetic field.

near the North Pole. Charged particles from the solar wind collect in two belts around the earth, known as the Van Allen radiation belts. This zone of trapped particles extends from several hundred to 40,000 miles above the earth's surface.

Excess particles exit the radiation belt and flow along the earth's magnetic field lines. These lines direct the particles toward the poles. As the particles enter the earth's atmosphere, they collide with gas molecules, causing the molecules to release energy. This released energy appears luminous and is the cause of the aurora. The aurora occurs generally between 50 and 500 miles above the earth's surface, far above the height where clouds form and where most weather occurs. Since the earth's magnetic field lines enter and exit the atmosphere at high latitudes, this is where most auroral displays occur. The colors of the aurora depend on the molecular structure of the energized gases. Oxygen emits both red and green light, depending on the altitude. Nitrogen emits red and violet light.

AURORAS AND SUNSPOTS

Auroral activity is closely tied with the cycle of sunspots. Sunspots are cooler regions of the sun surrounded by extremely bright rings, and the frequency of their appearance reaches a maximum approximately every eleven years. The sun emits more radiation when sunspot activity is at a maximum, so the intensity of the aurora is higher during these periods. Solar flares, huge eruptions of solar gas, can also intensify the solar wind and lead to bright auroral displays. At their most intense, these solar flares can disrupt radio and satellite communications by sending large amounts of particles into the earth's upper atmosphere.
See also: Atmosphere; Radiation; Satellite; Wind.

Barometer

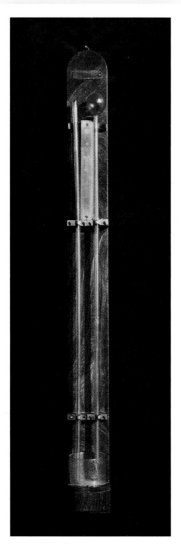

The first barometer, invented by Evangelista Torricelli in the 17th century, was much larger than those used today. Pictured is a reproduction of a 17th century barometer.

Invented in 1643 by Evangelista Torricelli, a barometer is an instrument used to measure the atmospheric pressure. The atmospheric pressure is the force per unit area exerted by moving air molecules. Knowing the air pressure is important, since changes in air pressure can signal changing weather. Meteorologists use maps of plotted barometric readings to locate positions of weather systems moving around the earth.

The first barometer was a mercury barometer, which consists of a long glass tube with a closed top end that is set in a dish of mercury. In a mercury barometer, the weight of air on the mercury in the dish forces the mercury to rise inside the tube. Air is removed from the top of the tube, above the column of mercury, allowing it to rise and fall freely. The height it reaches, in inches, is a measure of the magnitude of air pressure and is read as "inches of mercury," or just "inches."

Another kind of barometer, more commonly used today, is the aneroid barometer. This instrument consists of a small, flexible metal container called an aneroid cell, partially evacuated of air. Changes in air pressure result in expansion or contraction of the cell. The action on the cell is modulated by a spring and connected to an indicating arm. The size of the cell is calibrated so that the moving arm correctly points to the atmospheric pressure indicated on the barometer's face.

A pen can be attached to the moving arm of an aneroid barometer so that it traces the air pressure on a slowly rotating drum. This instrument is called a barograph, and meteorologists use it to record air pressure over time. A computer connected to an aneroid barometer can electronically record a barograph trace. Since atmospheric pressure decreases with increasing height in a predictable way, a barometer can also be calibrated to measure altitude.

When Torricelli first used his mercury barometer, he observed the liquid to reach a height of around thirty inches inside the glass tube. This is a typical reading for sea level atmospheric pressure. He discovered that mercury was a better liquid to use in a barometer than water. In the same tube, much lighter water rises over thirty feet where mercury rises only thirty inches.

Lower atmospheric pressures are associated with cyclonic storms, while higher pressures usually accompany fair weather systems. A barometer might read 28.75 inches in an intense low-pressure area, while in a strong high-pressure system it might indicate 30.75 inches. The lowest pressure ever recorded was 25.70 inches of mercury in a Pacific typhoon. The highest pressure ever recorded was 32.01 inches in Siberia during clear, cold conditions in winter. *See also: Air; High Pressure; Low Pressure; Weather.*

BLOCKING PATTERN

Normally, weather systems over middle and high latitudes move from west to east. Meteorologists use the term "blocking pattern" to describe the situation when an upper-level weather system stalls, thereby blocking the normal west to east progression of weather systems. Blocking patterns can result in the persistence of similar weather over a certain region for days, or even weeks. The extended periods of dry or wet weather that lead to droughts and floods are often caused by blocking patterns.

Meteorologists observe two distinct types of blocking patterns. The first, called a Rex block, was named after Daniel F. Rex, the meteorologist who studied it in the early 1950s. A Rex block exists when the jet stream, a river of fast winds in upper levels of the atmosphere, encounters a stalled ridge and splits into two branches. One branch of the jet stream travels poleward around the upper-level ridge, while the other travels equatorward. The second kind of blocking pattern is called an omega block, so called because upper-level wind flow takes on the appearance of the Greek letter omega: Ω. With an omega block, the jet stream simply diverts poleward around a ridge rather than splitting into two branches.

When a blocking pattern is present in the atmosphere, similar weather can occur day after day over the same areas in the vicinity of the block. A blocking pattern played a role in the disastrous Mississippi River flood of 1993. In this case, a blocking ridge over the North Pacific forced jet stream winds to dive southward into a trough over the western United States. Flooding rains ensued over the Midwest as disturbances rippling out of the trough encountered abundant atmospheric moisture over the Midwest.

Air generally flows from west to east around the earth, but it does so in a wavy pattern. These waves are associated with the movement of large masses of warm and cold air around the earth, and they extend for hundreds or even thousands of miles from crest to crest. A wave with relatively warm air and high air pressure in upper levels of the atmosphere is called a ridge. Air flows in a clockwise, or anticyclonic, direction around a ridge. A wave with low air pressure and cold air aloft is called a trough. Wind blows in a counterclockwise, or cyclonic, direction around a trough.

Sometimes one wave can overtake or merge with another wave. Depending on whether the waves are ridges or troughs, one can cancel the other out, or they can combine and amplify into a much stronger wave. For example, a trough moving into a large ridge might weaken, while a trough overtaking another trough might form a very strong, or "deep," trough. Blocking can occur when atmospheric waves combine and increase in amplitude. Larger waves move much slower and sometimes stall, producing a blocking pattern.

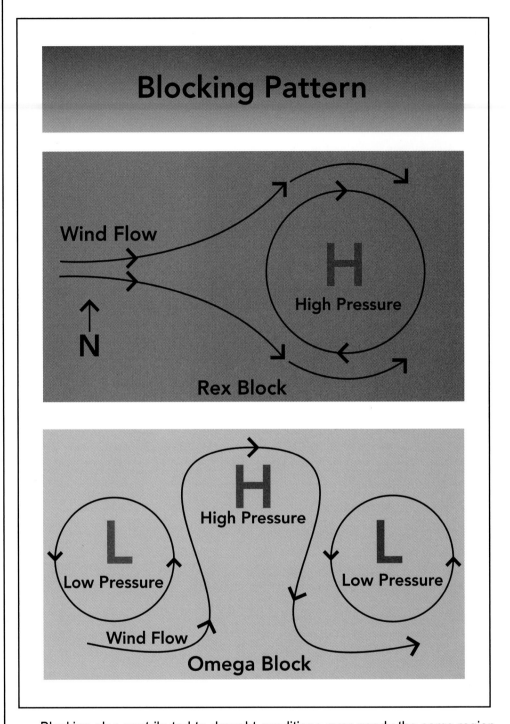

Blocking patterns form when areas of high and low pressure stall, forcing other weather systems to move around them.

Blocking also contributed to drought conditions over nearly the same region in 1988, when Mississippi River levels dropped to all-time lows. This time the blocking ridge was centered directly over the central United States. Warm air aloft associated with the ridge resulted in a stable atmosphere. A stable atmosphere is one that prevents the air motions that give rise to clouds and rainfall. This dry, stable pattern persisted for weeks, leading to drought conditions. *See also: Cloud Formation; Drought; Floods; Jet Stream; Ridge; Trough.*

Butterfly Effect

The butterfly effect states that the flapping of a butterfly's wings in China can cause a tornado in Texas weeks later. Rather than being a provable theory to be taken literally, it is more an illustration of how the atmosphere behaves. The term "butterfly effect" is attributed to Edward Lorenz, a professor at the Massachusetts Institute of Technology, who was one of the first to use computers to make crude weather predictions. The consequence of the butterfly effect is that the atmosphere is essentially unpredictable over long time periods.

In the early 1960s, Lorenz programmed a computer to make basic weather predictions. To do this, the computer was given a set of initial conditions. These initial conditions described the temperature and other variables of the atmosphere over a given time period. The computer would then calculate the value of the initial conditions at a specific time in the future, using complex mathematical equations. This process was repeated over and over, each time inserting the output of the previous calculations back into the equations to compute the weather farther and farther out in time.

What Lorenz discovered is that seemingly inconsequential variations in the initial conditions could result in very large differences in the predicted weather. With each cycle of calculations, small differences in the atmospheric variables would grow in magnitude. Lorenz could start with two sets of initial atmospheric variables only 0.001 percent different from each other and end up with radically different calculations of the future state of the atmosphere.

Therefore, a small change in the state of the weather might grow into a much larger change over time. The butterfly effect describes how the tiny air currents generated by the flap of a butterfly's wings might affect slightly larger air currents around the butterfly. These air currents in turn would alter larger air currents, and so on, in a cascading chain of cause and effect. The result, according to this theory, might be a weather system that could spawn tornado-producing thunderstorms over Texas two weeks after the butterfly flaps its wings in China.

The butterfly effect has important consequences for weather prediction. Weather is a result of the motions and physical properties of the vast number of air molecules in the atmosphere. Modern computers use complex mathematical equations to calculate the temperature, pressure, and movement of these molecules not only on the surface of the earth but thousands of feet up in the atmosphere. These computer models are not nearly powerful enough, however, to calculate the motion of every air molecule in the atmosphere.

The most widely used model today calculates weather on a grid of points superimposed over North America. The points on the grid are spaced 32 kilometers apart at forty-five different levels of the atmosphere. Thus, the computer knows the state of the weather only at each point, but not in between. Small variations in the weather in between these points can lead to drastic differences between what the computer predicts and what actually happens over time. Studies have shown that because of the butterfly effect, detailed weather forecasts (such as reliable temperature and rain or snowfall predictions) may never be feasible beyond a period of weeks. *See also: Air; Atmosphere; Rain; Temperature; Weather.*

CATALINA EDDY

A Catalina eddy is a small-scale circulation in the atmosphere that forms along the southern California coast. It is named after a small island in the Pacific, west of Los Angeles. A Catalina eddy can cause widespread low clouds and fog across coastal southern California, including the Los Angeles area. Sometimes the circulation is strong enough that the clouds produce drizzle or rain, even in the middle of the summer dry season when no fronts or storm systems are nearby.

CAUSES OF A CATALINA EDDY

A Catalina eddy occurs when air flowing from the north interacts with the unique topography of coastal southern California. In general, the coast of southern California is oriented from northwest to southeast. North of Los Angeles the land juts out into the Pacific at a location called Point Conception. Southward from there the coastline turns inward toward the east in a long curve that extends southward to the Mexican border. Just inland and more or less parallel to the coast are mountains. Air flowing from the north along the coastline rounds Point Conception and bends to the east. The mountains and the naturally curving shape of the coast cause this bending air to spin up into a counterclockwise circulation, known as a Catalina eddy.

The circulation of a Catalina eddy spreads moist air from over the ocean inland where it is blocked by the mountains. This moisture can form clouds and fog. The stronger the circulation is, the more moisture flows into coastal southern California. This layer of moist, ocean air is known as the marine layer, and the deeper it is, the thicker the clouds become. Typically the afternoon sunshine burns these clouds away. However, if a Catalina eddy becomes particularly

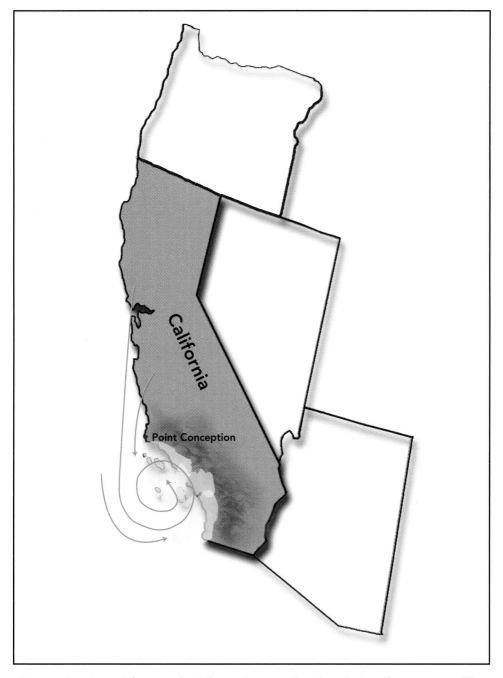

A Catalina eddy can form when northerly winds blow along the California coast.

strong, clouds and fog can last through an entire day during the summer. They can even produce drizzle or rain, an unusual phenomena in this relatively dry part of the country during summertime.

How Catalina Eddys Are Detected

Meteorologists look for a Catalina eddy using satellite photographs and wind readings from weather stations along the coast. Local forecasters know that a Catalina eddy will affect cloud cover and therefore temperature over a large, densely populated area. When they see that one has formed, they adjust their forecasts accordingly. The Catalina eddy is one example of how local terrain can result in unique weather conditions, and their detection shows how meteorologists use their knowledge of local weather to help them make better forecasts. *See also: Air; Atmosphere; Cloud Formation; Fog; Oceans; Rain; Satellite; Temperature; Wind.*

Climate

The climate of a location is determined by averaging its temperature and precipitation over many years. While extreme cold, heat, rain, and snow occur from time to time at any given location, the average weather usually remains consistent from year to year. Over centuries, however, natural processes or human influences can slowly change the climate of a location. How the temperature and precipitation change over time has a significant impact on agriculture and human settlement.

Volcanoes are linked to climate change.

Geographic Influences

The earth's climate zones have been shaped gradually over millions of years by the natural process of plate tectonics. This term refers to the slow movement of large sections of the earth's crust, known as plates. As these plates move, they change the distribution of land masses and oceans on a global scale. Since global temperature, air pressure, and wind patterns are influenced by the arrangement of oceans and continents, plate tectonics can significantly alter the climate of a region over millions of years. Scientists have speculated that Antarctica, now covered in snow and ice, was once much closer to the equator with a climate that supported abundant plant and early animal life.

Volcanic activity has also been linked to climate change, especially over shorter time periods. Large volcanic eruptions can spew huge amounts of ash, gas, and dust high into the atmosphere. Winds at high altitudes blow this volcanic output around the earth, forming an extremely thin layer of microscopic particles that can block a small percentage of the incoming solar energy. This nearly imperceptible decrease in sunlight is thought to result in temperature reductions of a few tenths of a degree. These kinds of small changes in weather can escalate into larger variations in regional climates over a period of months and years.

Changes in Solar Energy

Climate is also affected by changes in the earth's orbit around the sun over tens of thousands of years. These orbital changes include slight variations in the shape of the path that the earth takes around the sun, as well as changes in the tilt and wobble of the earth's axis. These orbital variations affect how much solar

The belief that changes in the earth's orbit have an influence on the earth's climate is known as the Milankovitch theory. Climatologists have attempted to link the Milankovitch theory with the spreading and contracting of global ice sheets many thousands of years ago. They believe that changes in the orbit of the earth reduced the amount of solar energy received in the Northern Hemisphere. This caused the ice and snow that was building up over thousands of years in the polar regions to spread southward, resulting in an ice age. Because orbital changes are cyclical, eventually the earth transitioned back into a warmer climate and the ice sheets receded, again over many thousands of years.

Climate is directly influenced by the way in which sunlight strikes the earth.

energy the earth receives between summer and winter, and between Northern and Southern Hemispheres. Variations in solar energy are directly related to temperature and pressure, which in turn affect wind and precipitation patterns.

Another way climate can change is through variations in the sun's output of solar energy. The amount of energy emitted by the sun changes according to a process known as the sunspot cycle. Sunspots appear as large dark areas on the surface of the sun. While these dark areas are cooler than the rest of the sun's surface, they are ringed by extremely bright areas called faculae. The heat of the faculae more than compensates for the cooler regions within the sunspot, so periods of high sunspot activity correlate to high solar energy output and higher temperatures.

Human Influences

Humans also affect climate. Industrial emissions, primarily from the burning of fossil fuels like coal and oil, increase the amount of carbon dioxide in the air. This gas, known as a greenhouse gas, is an efficient absorber of outgoing radiation from the earth. The more carbon dioxide in the atmosphere, the more radiation is absorbed and the warmer it becomes. The amount of warming that results from increased greenhouse gases is a source of great speculation and controversy, with estimates ranging from a few tenths of a degree to several degrees over a period of decades.

Emissions from industry and vehicles can cause pollution to gather over cities.

Climate change can occur not only on a global scale but also over much smaller areas. The urban heat island effect is the term given to describe how cities affect local temperatures. As urban areas grow, buildings and roads replace trees and other vegetation. Concrete and pavement absorb much more solar energy than forests and fields, so cities and surrounding locations become warmer than outlying rural areas. This can be particularly noticeable at night, as a city only slowly radiates the heat it gained during the day. Nighttime differences in temperature between a city and a rural location only fifty miles away can be ten degrees or more. *See also: Air; Atmosphere; Oceans; Precipitation; Rain; Radiation; Snow; Temperature; Urban Heat Island; Weather; Wind.*

CLIMATE CHANGE

Climate is the long-term average of weather conditions at a given location. Meteorologists classify areas with similar climates into climate zones.

CLIMATE CONTROLS

The factors that influence the prevailing weather in an area over a period of years are called climate controls, and they are closely linked to the region's location, geography, and global wind patterns.

One climate control is sunshine, which varies with latitude and season. Since the sun shines on lower latitudes at a much more direct angle than at high latitudes, regions closer to the equator receive much more energy from the sun and have warmer climates. Sunlight also varies with season. The changing seasons are caused by the tilt of the earth's axis as it rotates around the sun. In January, the Northern Hemisphere is pointed away from the sun, causing the sun to be lower in the sky than in summer months. Areas near the North Pole have little or no sunshine for weeks at a time. This causes their already cold climate to become colder in winter.

The distribution of land masses and oceans across the globe also influences climate. In general, land areas gain and lose heat much more quickly than water. Therefore, land areas can become much hotter or colder than areas over the ocean. Islands and areas near the coastline of a continent often have milder climates than locations near the interior of continents. Ocean currents also modify the climate of some locations. The air above these currents takes on the temperature characteristics of the water and the movement of these currents can transport warm and cold temperatures over thousands of miles. Ocean currents also affect precipitation. Areas adjacent to cold ocean currents are often drier than areas adjacent to warm currents. This is because colder air near the earth's surface is denser and less likely to rise and form clouds, rain, and snow.

A third climatic influence is the circulation of air around the planet. In general, air rises at the equators and flows northward at high altitudes. This rising air carries humid tropical air skyward, forming clouds that produce showers and thunderstorms. This results in a wet climate for many locations in the tropics. The northward-moving air tends to sink back to the ground at about 30° north latitude. Sinking air causes moisture to evaporate, suppressing clouds and rainfall. This explains why many of the world's desert climates are near this latitude. Another zone of sinking air lies at high latitudes, causing relatively dry weather near the poles. From 30° to 60° north latitude, the wind flow is generally westerly and is highly variable. In this region, air circulation has little effect on the climate because the masses of cold and warm air mix and spread.

CLIMATE ZONES

One of the most widely used systems to classify climate zones is the Köppen system. This method uses long-term weather averages to categorize regions into five major zones, each with subdivisions that are based on seasonal variations of temperature and precipitation.

The first Köppen zone is the tropical zone. The average temperature for all months in a Köppen tropical zone must be greater than 64°F. Various tropical subregions are divided according to whether the area has a dry season or not. Rainforests, such as those in the Amazon basin in Brazil, are located in a tropical climatic zone.

The second kind of climatic zone is a dry climate. In a dry climate, water evaporation exceeds precipitation. Evaporation, the process whereby water changes from liquid to gaseous form, depends on sunlight and temperature. The deserts of the world are classified as dry climates, including the Sahara, parts of the southwestern United States, and central Australia. Another kind of dry climate is a semiarid grassy plain, known in some parts of the world as a steppe. The Great Plains of the United States is an example of this kind of semiarid, dry climate.

The third kind of climate is called "continental with mild winters." In these locations, the average temperature of the coldest months must be between 27°F and 64°F. Deep snowfall and bitter cold are rare in these parts of the world, while summers can be hot. Examples of this kind of climate include the southeastern United States, eastern China, and southeastern Australia. A special climatic subregion is called a Mediterranean climate. Parts of the world with this kind of climate, including coastal California and areas around the Mediterranean Sea, have mild winters and dry summers.

The fourth climate category is called "continental with severe winters." Locations in this climatic zone have average temperatures that are less than 27°F in the coldest months, though summers can be warm. Winters are characterized by snowstorms and occasional episodes of subzero temperatures. Parts of Russia, north central United States, and Canada are examples of this kind of climatic zone.

The last kind of climate is a polar climate. In these locations, the average temperature of the warmest months is no greater than 50°F. Precipitation is variable, though often on the low side. *See also: Air; Cloud Formation; Oceans; Precipitation; Rain; Seasons; Snow; Temperature; Thunderstorms; Weather; Wind.*

Cloud Classification

Clouds are classified according to the altitude of their base and are named with descriptive Latin terms. Most clouds develop in the lowest seven miles of the atmosphere, though some can form at extremely high levels. They take a variety of shapes and sizes, depending on how they form and at what temperature.

Lenticular clouds get their name from their smooth, lens-like shape. These clouds are sometimes mistaken for alien "flying saucer" spacecraft.

Weather observers describe the sky by the fractional extent of cloud cover. A clear sky is one where less than 1/10 of the sky contains clouds. A scattered sky has between 1/10 and 5/10 cloud cover, and a broken sky between 6/10 and 9/10. Above 9/10 the sky is described as overcast. When the sky cover is broken or overcast, the lowest layer of clouds is called the ceiling.

High clouds, including cirrus, cirrostratus, and cirrocumulus, generally form at altitudes from 16,000 to 45,000 feet, depending on season and latitude. The farther north or colder the atmosphere, the lower these kinds of clouds can form. They consist mainly of ice crystals and often have a thin, wispy appearance as they are blown by high-level winds.

Middle clouds typically form between 7,000 and 23,000 feet and contain mostly water droplets. They include varieties such as altostratus, which appear as a solid layer of clouds through which the sun may be dimly visible. Another example of a middle cloud is altocumulus, which sometimes takes the appearance of cotton balls.

Clouds that form at low elevations are typically the clouds that produce precipitation, since the densest concentration of atmospheric moisture exists at low levels of the atmosphere. They form anywhere from 7,000 feet down to the earth's surface. The gray and sometimes drizzly stratus is a low cloud, as is the dark, rainy, and ragged nimbostratus. Low clouds contain mainly water droplets, but in winter they can also carry ice crystals and snow.

Clouds that show significant vertical development are categorized separately. These include the puffy cumulus clouds, which often dot the sky on a summer day, and the towering cumulonimbus, or thunderstorm cloud, which can rise over 60,000 feet high. Both clouds contain mostly water droplets, though the tops of cumulonimbus clouds consist mainly of ice crystals that spread downwind of the main cloud. This part of the cumulonimbus is called the anvil, owing to its flat appearance.

Meteorologists sometimes use descriptive words to help further distinguish certain clouds. Fractus is a word usually applied to cumulus clouds with a ragged, fractured appearance. Humilis is a descriptive term indicating vertical development in cumulus clouds, while castellanus describes altocumulus clouds when they have a marked towerlike appearance. Lenticularis is a word used when clouds such as cumulus and cirrocumulus develop in a lenslike shape. These clouds are usually formed in the smooth, fast flow of air downwind of a mountain. *See also: Air; Atmosphere; Cloud Formation; Precipitation; Snow; Temperature; Thunderstorms.*

Cloud Formation

Clouds form in the sky when the invisible water vapor in air condenses into tiny, visible water droplets or freezes into tiny ice crystals. Clouds not only can produce precipitation, but they also regulate the surface air temperature by

Cloud Formation

preventing sunlight from reaching the earth. They are part of the hydrologic cycle, the process whereby liquid water and water vapor circulate between the oceans, lakes, rivers, and atmosphere.

Most clouds form when vertical currents of air rise and the air cools. There are several processes that cause vertical air currents. The first is convection, a process responsible for puffy cumulus and thundering cumulonimbus clouds. These clouds form when sunshine warms the ground, which warms the air in contact with it. This warmed air then rises and cools. Another way that air can rise is when large hills and mountains force flowing air upward. A third process that causes rising currents is when air converges, or comes together, near the surface of the earth. As air comes together, it is forced upward. Likewise, when air diverges, or spreads apart, at high levels of the atmosphere, air from below rises to take its place. Finally, fronts can force air to rise in large quantities over thousands of miles.

Once the air is higher in the atmosphere, it begins to cool. As the air cools, the water vapor molecules in the air move slower and are more likely to stick together when they collide. As water vapor molecules stick together, they form microscopic water droplets. The process whereby water vapor turns into liquid water droplets is known as condensation. Evaporation occurs when liquid water molecules turn into gaseous water vapor. The temperature at which condensation occurs more frequently than evaporation is known as the dew point temperature. When rising air cools to the dew point temperature, condensation exceeds evaporation and a cloud forms.

Some clouds form under unique circumstances. Mammatus clouds form in the descending air underneath the flat top, or anvil, of a cumulonimbus cloud and have a distinct udderlike appearance. Contrails are clouds that form when exhaust from jet airplanes mixes with very cold air at high altitudes. Banner clouds form as air crosses over high mountain peaks and is forced upward into a cloud that trails downwind from the mountain like a banner. Some of the rarest and highest clouds are called noctilucent clouds. These clouds develop in the mesosphere, nearly fifty miles high, as ice crystals form on meteoric dust particles.

Cumulus clouds form when sunshine warms the ground, causing air to rise into the atmosphere.

43

Airborne microscopic particles serve as a catalyst for condensation and cloud formation. These particles, called condensation nuclei, include smoke, dust, sea salt, and pollen. Condensation nuclei range in size from less than 0.2 micrometers to around 1 micrometer and exist in concentrations of 100 to 1,000 per cubic centimeter of air. Some nuclei, such as microscopic sea salt particles, are called hygroscopic because they attract water vapor and are more favorable for cloud droplet formation. When a hygroscopic nucleus dissolves in a tiny cloud droplet, some of the water-attracting nuclei molecules move to the surface of the drop where they attract more water vapor. This enables the droplet to grow larger.

Meteorologists have found that cloud droplets remain in liquid state at temperatures as low as -30°C. Liquid water that exists below freezing is referred to as supercooled water. Even clouds that produce snow in the middle of winter consist mostly of supercooled water droplets. This is because the molecular motions in extremely small water droplets prevent the droplets from freezing. For a droplet to freeze into ice, a certain amount of water molecules must lock into a rigid crystalline structure. In extremely small cloud droplets, fewer molecules are available to do this and they are constantly jostling around, preventing freezing from taking place. Another reason why the droplets don't freeze is that nuclei favorable for the formation of supercooled water droplets are much more abundant in the atmosphere than those favorable for the growth of ice crystals. Whether the nuclei cause the formation of water droplets or ice crystals depends on their molecular structure. If their molecular pattern is similar to that of ice, then ice crystals are more likely to form. *See also: Air; Atmosphere; Cloud Classification; Dew Point; Oceans; Precipitation; Snow; Temperature; Water Vapor.*

COLD FRONT

A cold front marks the advancing edge of a relatively colder and drier mass of air. On a weather map, it appears as a blue line with triangles pointing in the direction the front is moving. A cold front can be hundreds of miles long and extend thousands of feet into the atmosphere.

Even though a front appears on a weather map as a distinct line, in the atmosphere a front is actually a much wider zone of transition from a lower to a higher molecular density of air. A strong cold front can be narrow and abrupt, bringing a rapid change from warm to cold air as it passes. A weaker cold front can be over a hundred miles in width, with a gradual change in weather from the warmer to the cooler side of the front. Cold fronts do not extend straight upward

Cold Front

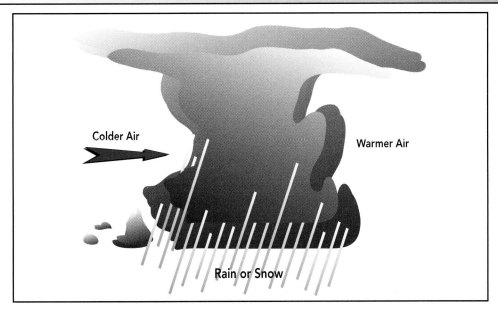

Colder air wedges into warmer air along a cold front, causing clouds, and rain or snow.

into the atmosphere; rather, they slope backward over the advancing cold air. This is because colder, drier air is usually denser than the air it is displacing. Therefore, the colder air wedges underneath the warm air as it advances.

This wedging action often results in strong upward vertical motions in the atmosphere along and ahead of the front. As air rises, it cools and expands, allowing water vapor in the air to condense into clouds that can produce rain or snow. If enough moisture is present ahead of the front, or if the vertical motions are strong enough, the weather along a cold front can become very active. Heavy showers of rain or snow with gusty winds often mark a strong cold front as it passes. In spring and summer, strong thunderstorms can form along cold fronts. In winter, cold fronts can bring rain or snow, followed by sharply falling temperatures.

Not all fronts are active, however; the passage of a weak cold front can be barely noticeable in the summertime. Temperatures sometimes differ by only a few degrees across a weak summertime cold front, with the air just slightly less humid after the front passes. Other cold fronts are completely devoid of clouds or precipitation, owing to a lack of moisture in the atmosphere.

Cold fronts typically travel at 15 to 25 mph and usually move from west to east, northwest to southeast, or north to south. Occasionally a cold front can move from east to west, as sometimes happens in New England. Here, a counterclockwise flow around a high-pressure system in the North Atlantic can drive cold

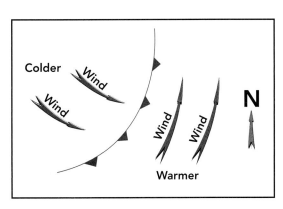

A cold front marks the leading edge of advancing cold air.

45

air into New England from the ocean. *See also: Air; Atmosphere; Cloud Formation; High Pressure; Precipitation; Water Vapor; Weather Maps.*

COLD WAVE

A prolonged period of temperatures well below average for the season is referred to as a cold wave. These episodes typically occur during winter when cold air is more widespread. Cold waves can threaten crops and transportation and can pose serious risks to health.

In North America, cold waves are often the result of large areas of frigid air gathering over northern Canada and Alaska. Temperatures in these regions can drop as low as -60°F as long, dark nights and snow cover allow cold air to build up over time. Since cold air is very dense, large cold air masses result in high surface air pressures. As a large amount of air gathers over a region, it eventually will begin to spread southward.

The cold air moves as the result of two processes. The force of air pressure pushing down and outward from the center of the cold, dense air mass causes the edges of the air mass to spread outward. Winds in the upper levels of the atmosphere also contribute to the movement of large cold air masses. High-level winds that converge, or come together, add air over a region and cause surface air pressure to rise. As the zones of upper-level convergence move along, so do the areas of surface high pressure and their accompanying cold waves underneath them.

Dangers associated with cold waves include frostbite, which is the freezing of body tissue, and hypothermia, the condition in which a person's body temperature drops below 95°F. Both of these events occur after prolonged exposure to cold air, or if a person is not well protected against the cold. As cold waves spread south, normally milder regions in the southern United States can experience damage to crops and fruit trees. Rivers that freeze during a cold wave can halt commercial barge traffic; snow and ice that sometimes accompany the leading edge of a cold wave often result in extremely dangerous driving conditions.

North America usually experiences several severe cold waves each winter. These events distinguish themselves from the normally colder winter weather by how much colder the temperature is from the average. In 1996, for example, a severe cold wave spread across much of the United States in late January and early February. Temperatures of thirty degrees below normal were common across much of the Midwest. The temperature in Minneapolis, Minnesota, dropped below -20°F for six consecutive nights. A temperature of -60°F was recorded at Tower, Minnesota, setting an all-time state record low temperature. The cold spread into the South, where temperatures fell into the single digits. All in all, over 400 record temperatures were set across the country in the cold wave. At least 100 people died as a result of the cold, and crops in the southern United States suffered between $50 and $60 million in damage. *See also: Air; Air Mass; Atmosphere; Frostbite; High Pressure; Hypothermia; Snow; Temperature.*

CONVERGENCE

The flow of air in a region of the atmosphere is said to be convergent when more air enters the area than is able to exit. This accumulation of air often results when winds from different directions blow against each other. Meteorologists look for areas of convergence to help track and predict large-scale weather systems as well as small-scale areas of precipitation. Convergence can cause clouds and precipitation when it occurs near the earth's surface, and it plays a role in the development of high-pressure systems when present at high altitudes.

The opposite of convergence is divergence, the process whereby the flow of air in a region spreads apart. Convergence and divergence often exist in tandem at different elevations in the atmosphere over the same region. For example, when

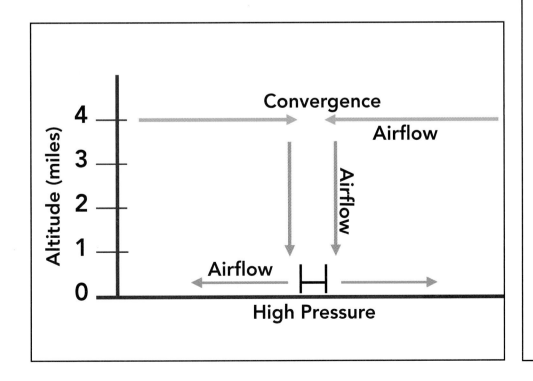

When convergence occurs at high altitudes in the atmosphere, air pressure often rises at the earth's surface.

two horizontal streams of air converge at ground level, the air is forced upward into the atmosphere. When this rising air reaches high altitudes, it encounters a stable layer of the atmosphere called the stratosphere. The stratosphere is described as stable because it prevents air from rising. When air rising from below hits the stratosphere, it is forced to diverge and spread apart horizontally.

One of the reasons why meteorologists are concerned with convergence and divergence is their influence on atmospheric pressure. The pressure of a given part of the atmosphere is determined by the amount of air molecules it contains. The amount of air molecules over a given area increases if convergence at one altitude in the atmosphere brings together more air than divergence removes from another. Air pressure increases as converging winds bring more air over an area. Thus, convergence can cause areas of high pressure to form and strengthen.

Air that flows together does not always converge, however. The key to whether an air flow is convergent or not is whether it increases the amount of air over a given area. Sometimes, air flowing together at an angle will speed up, decreasing the amount of air. Meteorologists use the term "confluence" to describe this situation. Convergence and confluence look very similar on weather maps and are easily mistaken for one another. ***See also: Air; Atmosphere; Cloud Formation; Divergence; High Pressure; Precipitation; Weather Maps.***

Coriolis Effect

The Coriolis effect (also referred to as the Coriolis force) is the phenomenon whereby the path of a moving object appears to bend toward the right in the Northern Hemisphere and to the left in the Southern Hemisphere. It is caused by the earth's rotation and acts on objects not connected with the earth, such as moving air, airplanes, and water. It alters the path of moving objects regardless of their direction, but significant displacement due to the Coriolis effect occurs on a time scale of hours rather than minutes. Meteorologists are concerned about the Coriolis effect because it modifies the wind flow of large-scale weather systems.

As the earth rotates eastward, its velocity is highest at the equator and decreases approaching the poles. Thus, an object flying from the equator toward one of the poles will begin its path with a high eastward velocity. As it proceeds poleward, the speed of the earth's rotation below it becomes slower. As the difference between the eastward speed of the object in the air and that of the ground

Coriolis Effect

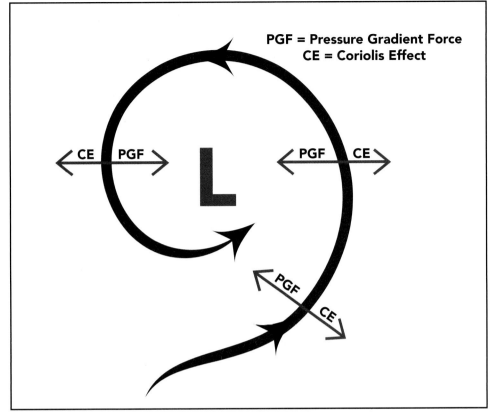

The Coriolis effect explains why air spirals in towards low pressure.

underneath it becomes greater and greater, the path of the object to a viewer on the ground will appear to curve increasingly to the east. In the Northern Hemisphere, this would look as though the path of the object was bending to the right. In the Southern Hemisphere, it would be seen as a bend to the left.

To compensate for the Coriolis effect, an eastbound airplane will fly to the left of its intended destination. As the plane progresses eastward, the earth rotates underneath the flight path. The destination city, therefore, moves eastward with the turning earth. From the plane's point of view, it also curves to the left as the earth turns on its axis. By the time the plane is ready to land, the earth will have spun the city to a point underneath the plane. To an observer on the ground, the plane will appear to curve to the right of its initial path.

One of the ways that the Coriolis effect influences weather is its capacity to cause a circulation of air around large weather systems. In general, air flows from high to low pressure. As it does, the Coriolis effect bends the flow of air to the right in the Northern Hemisphere. However, at the same time that the Coriolis effect acts to move air to the right, an even stronger force, called the pressure gradient force, or PGF, pulls air from the area of high pressure to the low-pressure area to the left. These conflicting forces cause the path of air in the Northern Hemisphere to spiral in a counterclockwise direction toward the center of the area of low pressure. Likewise, air flowing out from high pressure spirals in a clockwise direction.

Over a long duration, the Coriolis effect can similarly alter the direction of flowing water. However, the Coriolis effect is often mistakenly believed to be responsible for the direction that water circulates as it is drained. The Coriolis effect requires hours to exert a noticeable change in the direction of a moving object. Water drains from a sink too rapidly for the Coriolis effect to have any influence. The direction of draining water arises only from the shape of its container and the direction the water was moving before being drained. ***See also: Air; High Pressure; Low Pressure; Weather; Wind.***

CYCLOGENESIS

The formation and strengthening of a low-pressure area is called cyclogenesis. This process can be rapid and strong, producing a large and powerful storm system, or it can be slow and weak, resulting in a minor weather disturbance. Since low-pressure systems are usually accompanied by clouds and precipitation and can sometimes develop into large and dangerous storms, meteorologists closely watch for conditions that lead to cyclogenesis.

Cyclogenesis can be thought of as the consequence of a spreading, or divergent, flow of air that decreases the amount of air at high elevations in the atmosphere. As this air is removed, air rises from below to take its place. This causes winds near the earth's surface to come together, or converge, to take the place of the rising air. Over time the rotation of the earth imparts a counterclockwise rotation to the converging winds. If the diverging air aloft removes more air than can be replaced by converging air at the surface, then the overall air density will decrease and pressures will fall. The result is a low-pressure system.

Cyclogenesis can occur anywhere under the right conditions. It occurs more frequently, however, east of large north-south mountain ranges and along continental coastlines where warm ocean currents flow northward. Cyclogenesis tends to occur along fronts, since fronts are zones of relatively lower atmospheric pressure.

Meteorologists refer to the formation of low-pressure areas east of mountain ranges as "lee-side cyclogenesis." In these situations, as the westerly flow of air crosses the highest elevations of the mountain range, it tends to curve toward the southeast. As the air continues east of the mountain range and over lower elevations, it curves back to the northeast. Meteorologists refer to the curvature of the air flow, or its spin, as vorticity. As air flows over and down the east side of the mountain range, its vorticity increases. The counterclockwise, or cyclonic, spin that is imparted to the air forms the spinning circulation of a low-pressure area. These kinds of low-pressure systems often form in the Plains states just east of the Rocky Mountains. When lee-side cyclogenesis occurs together with diverging air aloft, the result can be a particularly strong area of low pressure.

Low-pressure systems also frequently form during winter where a warm ocean current runs northward along a coastline. The difference in temperature between the warm water and the cold air over the land can be a strong contrast. The relatively warmer ocean warms the air over it and the acceleration of the air molecules causes the air density to decrease and its pressure to fall. Areas of upper-level divergence that move over this zone of preexisting low pressure result in cyclogenesis. This process frequently occurs along the East Coast of

When strong cyclogenesis occurs along a coastline, the resulting storm can cause high winds and tidal flooding.

The Macmillan Encyclopedia of Weather

Low pressure areas, which develop in the process of cyclogenesis, are often accompanied by stormy weather.

the United States where the Gulf Stream travels northeastward off the North and South Carolina coast. Such coastal storms can be large and destructive, with high winds and heavy rain or snow. ***See also: Divergence; Low Pressure; Oceans; Precipitation; Temperature.***

Degree Day

A degree day is a numerical measurement of how warm or cold a season has been. People use degree days to calculate when crops should be harvested and how the daily temperatures over several months affect heating and cooling needs for homes and businesses. The term "day" in degree day is somewhat misleading, as degree days are not a unit of time but rather of temperature.

A heating degree day is a unit of measurement that relates how cold a winter season has been. The number of degree days for a given day is calculated by subtracting the mean temperature of that day from 65. Thus, if the mean temperature of a given day in January was 40°F in Denver, then Denver would gain 25 heating degree days for the winter heating season. Many degree days can accumulate in one calendar day. Each day's total is added to the last so that the running tally can be compared to the average. The lower the temperature, the more heating degree days. The more heating degree days, the more fuel is consumed through the heating of buildings.

A cooling degree day is just the opposite. It is the measure of how far the daily mean temperature exceeds 65°F. Cooling degree days are used in the summer to calculate the cost of energy consumption due to air conditioning. Yearly tabulations of heating and cooling degree days are used by building planners to account for heating and cooling system requirements.

Farmers and traders of agricultural commodities use growing degree days to help them understand how crops ripen over a season. A growing degree day is defined as the mean daily temperature minus the crop's base temperature. The base temperature varies from crop to crop and is defined as the minimum temperature in which that particular crop can grow. As growing degree days accumulate over the spring and summer months, farmers can calculate at what rate their crops are maturing and when harvest might occur. ***See also: Temperature.***

Derecho

A derecho [day-RAY-cho] is a fast-moving, long-lived group of thunderstorms that produces very high winds along its path. Derechos are most common across the Midwest at night during the late spring and summer months, though they have been known to occur in the South and the Northeast. They can produce winds to hurricane force (74 mph) or greater over a wide swath and can even spawn occasional tornadoes.

Derechos can cause high winds that down trees and power lines.

The term "derecho," from the Spanish meaning "straight ahead," was first conceived in 1883 by a weather researcher in Iowa named Gustavus Hinrichs. While most major storm damage in Iowa at that time was attributed to tornadoes, he noticed that some storms left behind debris blown in a straight line rather than a swirling pattern. He correctly theorized that this kind of wind damage was produced by violent, fast-moving thunderstorms rather than tornadoes.

Since then, research has shown that fifteen to twenty derechos occur in the United States each year. Derechos form from thunderstorms that develop in the heat of the afternoon and evening. Under the right conditions, these thunderstorms sometimes merge together as they move along in the late evening hours. Warm, moist air flowing up and into the storms from the south and east provides the necessary energy to keep the storms strong. Meanwhile, torrential rain falling out of the thunderstorms drags cooler air downward. This cooler, downrushing air spreads out upon hitting the ground. Sometimes, strong winds blow into the back side of the storm on its northern or western flank. These strong winds are forced to the ground by the downrushing air. As this wind energy hits the ground it causes violent gusts that can down trees and power lines. The high volume of air spreading out underneath the storm propels the leading edge of the storm forward at higher speeds, and derechos have been known to race along at over 60 mph.

The forward surge of air causes the area of thunderstorms to take on a bulge, or bow, at its leading edge. This distinctive appearance clearly shows up on radar. Derechos can be 75 to 300 miles across and sweep out a path many hundreds of miles long before they dissipate around sunrise. For a thunderstorm complex to be classified as a derecho, it must produce wind damage along 250 miles of its path.

One of the reasons derechos form most commonly over the Midwest is because of the unique geography in this part of the country. Tropical air flowing northward from the Gulf of Mexico has a relatively flat, unobstructed path into the Midwest. It is this tropical air on which derechos feed. During spring and summer, this tropical air is at its warmest and most humid.

Derechos are also mainly a nighttime phenomenon. This is because the southerly flow of air that provides heat and moisture to the derecho is strongest at night. During the day, the sun's energy heats the ground, causing turbulent eddies and currents of air that mix with and slow the southerly flow of tropical moisture. After sunset, however, these vertical currents cease and the southerly flow can speed northward several thousand feet above the ground, largely uninhibited by turbulence. As the sun rises the next day, eddies and turbulence once again develop and choke off the southerly current of air. Having lost its source of energy, the derecho dissipates into clouds and showers. Occasionally, a dissipating derecho will leave behind enough of a circulation that it can reform the next night, far from where it originated. *See also: Air; Hurricane; Rain; Thunderstorms; Tornadoes; Turbulence; Wind.*

DESERTIFICATION

Certain parts of the world are classified as semiarid. These regions receive relatively low amounts of annual rainfall and do not support the growth of much more than short, drought-resistant grasses. Fluctuations in climate and changes in land use brought on by humans can turn these regions into deserts. This process, known as desertification, has a major and sometimes life-threatening impact on the human populations that live in these areas.

There are several ways that a semiarid region can become desert over a period of years. One reason involves the location of the semiarid region, especially if it is located in a part of the world that experiences monsoons, or seasonal wind shifts. Local populations rely on these wind shifts, which often blow from the south in the summer and carry rain-producing moisture, to water their crops and replenish drinking-water supplies. For reasons that meteorologists don't yet fully understand, these seasonal winds can arrive late or cease to blow entirely for periods of years. When the monsoonal rains become sporadic or do not arrive at all for several years, the resulting drought can lead to desertification.

Sometimes the lack of rain can cause drought conditions to become worse in a process known as feedback. With decreased rainfall, the soil over a region becomes drier. Local vegetation normally adds water vapor to the air in a process known as transpiration, but as the soil becomes drier these plants die. The dry soil and fewer plants means the air becomes drier as well, so rainfall-producing clouds are less likely to form. Less frequent rain makes the ground become even drier, and desertification worsens.

Humans have a major influence on desertification through their land management policies. Too many cattle or sheep can lead to overgrazing, which strips the land of its grasses, causes the air to grow drier, and loosens the soil so that it blows away more easily. Poor farming practices can have the same effect; as fields are cultivated, trees and other natural windbreaks are removed, allowing wind to blow faster and erode the soil more easily. Poor irrigation can also dry out the land.

One example of a region that has experienced desertification is the Sahel region of Africa. This area is located between the Sahara desert, which extends across most of northern Africa, and the tropical jungles of central Africa. Depending on fluctuation of climate and land use, vast sections of the Sahel are prone to drought and desertification. The 1960s and 1970s were a particularly dry period in the Sahel, and thousands of humans and animals died as the Sahara spread southward into areas that traditionally had been used for farming and herding. Another example of desertification occurred in the Great Plains of the United States during the 1930s. This example of desertification, known as the Dust Bowl, happened when farmers removed areas of foliage that served as windbreaks. As the wind blew stronger and rain stopped falling, soil eroded and was swept into great dunes of sand that buried thousands of farms over several states. Many people were forced to give up farming and migrate to different parts of the country. Since then, farmers in the Plains states have learned to plant rows of trees and fences that slow the wind, and have developed effective irrigation techniques. ***See also: Air; Climate; Drought; Oceans; Rain; Water Vapor; Wind.***

Dew

Tiny water droplets that form on exposed outdoor surfaces are called dew. Dew forms when air cools, and therefore occurs mainly during the early morning hours as the temperature reaches its lowest level. Unlike frost, dew provides beneficial moisture to plants and grasses, especially in dry climates. In some desert locations, dew is the main source of water for both plants and insects.

How Dew Forms

Dew forms as the temperature of the ground falls after the loss of daytime heating. As the ground temperature drops, so does that of the air in contact with it. Water vapor molecules in the cooling air move at slower and slower speeds, becoming more likely to stick to each other or other objects when they collide. The point at which the water vapor molecules move slow enough that they adhere more often than they fly apart is known as the dew point. When the air reaches the dew point, water vapor in the air condenses into tiny water droplets. These water droplets form on grass, cars, and other surfaces.

Dew consists of tiny drops of water that often forms on grass and plants late at night or early in the morning.

Conditions for Dew Formation

Certain conditions favor the formation of dew. A nighttime sky that is cloud-free will allow greater surface cooling. Like all objects that have a temperature, clouds radiate energy. When clouds are present, the ground absorbs some of this energy and remains relatively warmer. The warmer the ground, the less the air in contact with it will be able to cool toward the dew point.

A calm night also aids in dew formation. Wind mixes cold air near the surface of the earth with relatively warmer air just above it. This keeps the air temperature slightly higher than when calm conditions allow cold air to settle near the ground. *See also: Air; Dew Point; Frost; Temperature; Water Vapor; Wind.*

Dew Point

The concept of dew point requires an understanding of water and how it behaves in the atmosphere. Water vapor molecules are always present in air. The speed at which they move depends on the temperature of the air. When gaseous water vapor molecules collide and adhere to each other or to another object, they form a tiny liquid water droplet. This process is called condensation. On the other hand, individual water molecules will separate from liquid water and become gaseous water vapor. This process is called evaporation. Both processes occur at the same time in air, with gaseous water vapor molecules condensing into liquid water and liquid water molecules evaporating into gaseous water vapor.

The dew point is a measure of how much water vapor is contained in a volume of air. Specifically, it is the temperature to which air must be cooled for the rate of condensation to exceed the rate of evaporation. Meteorologists prefer using dew point, rather than relative humidity, to measure how humid the air is. The dew point temperature also helps explain how rising air forms clouds.

The dew point temperature is the temperature to which a volume of air must cool for the rate of condensation to exceed the rate of evaporation. As long as the air temperature remains above the dew point temperature, vapor molecules will collide and adhere but very quickly fly apart again. This prevents water droplets (such as those that form clouds and fog) from forming. When the air temperature reaches the dew point temperature, the rate of condensation exceeds the rate of evaporation, and cloud or fog droplets can then develop.

The amount, or concentration, of water vapor molecules in the air also affects dew point. As the air temperature falls, water vapor molecules move slower. Therefore, the more humid the air is, the less the air will have to cool before condensation exceeds evaporation. That results in a higher dew point temperature.

Meteorologists look at the dew point temperature to see how moist the air is over a given region. Dew point temperatures of 60°F are a fairly moist reading, while a dew point of 70°F is humid. On rare occasions, dew points will approach 80°F in summer when tropical moisture is present; such readings are associated with the most oppressively muggy conditions. The dew point temperature is a standard part of the hourly weather observations that are taken at thousands of reporting sites around the world.

Meteorologists prefer using the dew point temperature to relative humidity as a way of measuring humidity because the relative humidity will change when the air temperature changes, even if the amount of moisture in the air remains the same. The dew point temperature, on the other hand, will stay the same when the air temperature changes, as long as the amount of moisture in the air remains consistent. The same relative humidity reading, therefore, will feel different at two different temperatures. For example, a relative humidity of 50 percent when the air temperature is 70°F feels comfortable. However, a relative humidity of 50 percent when the air temperature is 90°F feels uncomfortably humid.

Dew point is an important concept in understanding cloud formation. Clouds form when air rises. As air rises higher, it moves into lower and lower atmospheric pressure, since less and less air is weighing down from above. The rising air therefore expands, and molecules move farther apart and slow down. The slowing down of molecules means that the temperature falls, and when the temperature reaches the dew point, condensation occurs and a cloud forms. *See also: Air; Cloud Formation; Fog; Temperature; Water Vapor; Weather.*

DIVERGENCE

The flow of air in a region of the atmosphere is said to be divergent when more air exits the area than is able to enter. This depletion of air often results when wind flow spreads apart. Spreading flows of air often occur at high altitudes in the atmosphere when air decelerates. The process is similar to that which

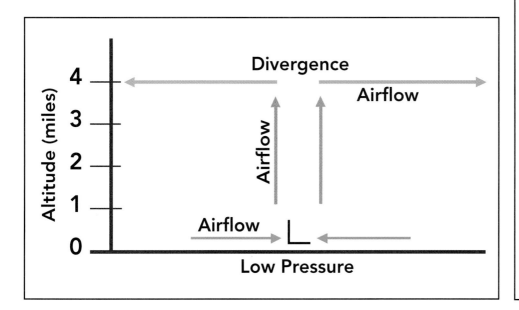

When divergence occurs at high altitudes in the atmosphere, air pressure often falls at the earth's surface.

occurs when a fast-moving, narrow river spreads out as it slows down. Meteorologists look for areas of divergence to help track and predict large-scale weather systems. Divergence plays a role in the development of low-pressure systems when present at high altitudes.

The opposite of divergence is convergence, the process whereby wind flows together and increases the amount of air over a region. Divergence and convergence often exist in tandem, at different elevations in the atmosphere over the same region. For example, when winds spread apart and remove air from high levels in the atmosphere, air is forced to rise from below to replace the loss of air above. To compensate for the rising air currents, more air is forced to spread in and converge horizontally near the earth's surface. Therefore, areas of divergence occur at high elevations over areas of convergence near the earth's surface, and vice versa.

One of the reasons why meteorologists are concerned with divergence is its influence on atmospheric pressure. The pressure of a given part of the atmosphere is determined by the amount of air molecules it contains. The amount of air molecules over a given area decreases, for example, if divergence at high altitudes in the atmosphere removes more air molecules than can be replaced by convergence at ground level. As the amount of air molecules decreases, the air pressure falls. Thus, divergence at high elevations in the atmosphere can cause areas of low pressure to form and strengthen.

Air that spreads apart does not always diverge, however. The key to whether an air flow is divergent or not is whether it decreases the amount of air over an area. Sometimes, air flowing apart at an angle will slow down. This can increase the amount of air over an area. Meteorologists use the term "diffluence" to describe this situation. Divergence and diffluence look very similar on weather maps and are easily mistaken for one another. ***See also: Air; Atmosphere; Convergence; Low Pressure; Weather Maps.***

DOPPLER RADAR

Radar, which stands for radio detection and ranging, is an electronic instrument that senses the location and intensity of precipitation. Doppler radar, in addition to sensing precipitation intensity and location, also detects wind direction and velocity. It therefore helps meteorologists visualize the structure of small-scale weather systems. Doppler radar has been particularly helpful in detecting circulations inside thunderstorms that lead to tornadoes.

THE DOPPLER EFFECT

Doppler radar depends on a phenomenon known as the Doppler effect. The Doppler effect relates the frequency at which energy bounces off an object with the movement of the object. A radar transmits pulses of microwave energy that bounce off of raindrops, snowflakes, and other small objects suspended in the air. The frequency at which the radar energy bounces back from the object depends on the object's movement. Objects moving away from the radar will reflect, or scatter, energy at a lower frequency. Objects moving toward the radar will scatter energy at a higher frequency.

A practical example of the Doppler effect occurs when a large truck passes by on a road. To an observer standing by the side of the road, the sound of the truck appears to start at a high pitch. The frequency at which sound waves approach the listener is higher since the truck is pushing the sound waves ahead of it. But as the truck passes by and travels away, the pitch falls. This is because the speed of the departing truck is subtracted from the speed of the sound waves, and their frequency therefore falls.

Doppler radar uses the Doppler effect to determine whether objects such as raindrops or snowflakes are blowing toward or away from the radar site. Understanding how airborne objects are moving results in an understanding of how the wind is blowing. On a radar screen, energy that has been scattered by objects blowing toward the radar is usually colored blue or green, while energy scattered by objects blowing away from the radar is colored orange and red. This data allows meteorologists to easily see where the wind is directed toward or away from the radar and how fast the wind is moving.

A Doppler radar is mounted on a steel tower and shielded from the elements by a protective dome.

HOW THE DOPPLER RADAR WORKS

Doppler radars have revealed the internal structure of very strong, highly organized thunderstorms, called supercells. As air rises into these storms, it rotates, taking a path that looks much like a corkscrew up through the storm. It is from this rotation, called a mesocyclone, that tornadoes sometimes develop. Doppler

Doppler radar can be used to scan thunderstorms for the swirling winds that may lead to a tornado.

radar can detect a thunderstorm's mesocyclone, the formation of which can precede a tornado by fifteen minutes or more. This allows meteorologists to issue timely warnings that can save lives. However, researchers have used Doppler radar to discover that mesocyclones form more frequently in thunderstorms than was previously thought. And more important, data from Doppler radars together with human observations have shown that fewer than 50 percent of these rotating thunderstorms go on to produce tornadoes. Meteorologists therefore require trained professional spotters and law enforcement to relay actual observations. These observations, together with data from Doppler radars, help weather forecasters issue more accurate warnings.

Doppler radar has been used for weather research for decades. A crude version of Doppler radar was first used to scan thunderstorms in Kansas in the late 1950s. It wasn't until the early 1970s, however, that computer technology allowed scientists to use Doppler radar to scan a tornado for the first time. After years of additional testing, the National Weather Service deployed a fleet of nearly 150 Doppler radars across the United States in the early 1990s. University and government researchers now drive portable Doppler radars close to tornado-producing thunderstorms in order to gain valuable data on how these storms develop. In May 1999, one such radar measured winds in excess of 300 mph in a violent tornado that struck Oklahoma City, Oklahoma. **See also:**

Mesocyclone; Precipitation; National Weather Service; Radar; Thunderstorms; Tornadoes; Weather; Wind.

DOWNBURST

A downburst is a violent rush of air about five miles in diameter that exits the base of a thunderstorm and spreads out upon hitting the earth. A more intense and localized variety of downburst is called a microburst, which concentrates its wind energy within a diameter of around 2.5 miles. Both downbursts and microbursts can generate wind gusts over hurricane force (74 mph), and some downburst-generated winds have been recorded at greater than 100 mph. In addition to downing trees and power lines, downbursts cause life-threatening danger to aircraft.

Downbursts occur when the normal circulation of air in a thunderstorm becomes intensified. In general, a thunderstorm consists of inflowing and outflowing air. The inflow carries warmth and moisture into the low levels of the thunderstorm. This moisture and warmth then rises high into the thunderstorm in a current of air called an updraft. The moisture contributes to rain, which falls to the ground, dragging air with it in another current of air called a downdraft. The heavier the rain, the more air is dragged toward the earth in the thunderstorm's downdraft. As the downdraft hits the earth, it spreads apart and flows out from underneath the storm. This outflow results in cool, gusty winds at the earth's surface that often precede the onset of rain by a minute or so as a thunderstorm moves over an area.

A downburst occurs when this downdraft strengthens far beyond its normal velocity. This can happen if an area of dry air several thousand feet in the atmosphere mixes with the downdraft. As rain falls into this drier air, some of the rain evaporates and the downdraft becomes drier. This drier air is more dense than moist air, so the downfalling air gains momentum. Also, the evaporation of rain removes heat energy from the air. Therefore, when rain evaporates, the air temperature around the rain falls. This cooler air adds to the density of the downrushing air. If enough dry air mixes with the downdraft, it can accelerate to a high velocity and become a downburst or microburst. The effect of a downburst can be likened to a stream of water splashing onto the ground and spreading apart.

Microbursts can be characterized as wet or dry, depending on whether they contain rain or not. In the western United States, microbursts are usually of the dry variety because the air is so much drier than areas east of the Rocky

Mountains. Thunderstorms that form in moist, tropical air in the southern states usually produce wet microbursts.

A meteorological professor and researcher named Theodore Fujita was the first to understand downbursts and how they can pose extreme danger to aircraft. Fujita studied airline disasters at John F. Kennedy Airport in 1975, near New Orleans in 1982, and in Dallas in 1985. Each of these incidents occurred as jet aircraft flew through or near downbursts. These accidents have led to the installation of wind shear warning systems at airports across the country as well as improved training for airline pilots. *See also: Air; Atmosphere; Theodore Fujita; Heat Transfer; Rain; Temperature; Thunderstorms; Wind; Wind Shear.*

DROUGHT

A drought is a period of abnormally dry weather, lasting months or even a year or more. Droughts are often accompanied by hotter than average weather. They can destroy crops, cause major water shortages, and affect commercial traffic on rivers.

The precipitation deficits that result in drought vary, depending on how much rain or snow a given location usually receives. For example, rainfall totals of a fraction of an inch over an entire summer in the central United States would lead to drought conditions, since these areas usually receive six to twelve inches of rain in this period. But only a fraction of an inch of rain over an entire summer in central California is close to what normally falls during this time period, and therefore would not result in a drought.

In the United States, drought conditions usually occur after weeks or months of below-normal rainfall or snowfall. Human activity can worsen the effects of a drought, depending on how fragile the local water supply is. Cities in the southwestern United States, for example, use large amounts of water in a part of the country that is normally dry, straining the naturally available water resources. Therefore, even slight or moderate drought conditions can impose the need for water restrictions.

Droughts can occur when weather systems stall or persist in the same pattern over long periods. For example, winds in the upper levels of the atmosphere may steer moisture-bearing storms away from a particular region for weeks or months. This kind of weather pattern occurs as large upper-level areas of high pressure become nearly stationary over a particular area. This kind of weather pattern is called a blocking pattern by meteorologists, since the upper-level area of high pressure blocks storms and forces them to travel around the high-pressure area.

Droughts result in the ground becoming dry and cracked. This piece of ground used to be the bottom of a pond.

If the blocking pattern lasts long enough, areas underneath the blocking high-pressure area will be deprived of moisture and may experience drought conditions.

The reasons why blocking patterns become established in some parts of the world during some seasons but not others are still not completely clear to meteorologists. Research has shown, however, that the movement of these large weather systems in upper levels of the atmosphere is related to the movement of large areas of warm and cold ocean water. The temperature of the ocean water can affect the temperature and pressure of the air above it. Over long periods of time, an area of above or below normal ocean water can result in large-scale areas of high and low pressure overhead, thereby altering the normal flow of weather patterns and sometimes producing the blocking pattern that can lead to drought.

Drought can be measured by a number of indices. These indices are computed using equations that relate precipitation, soil moisture, temperature, and other factors over weeks and months. One such index is called the Palmer

Drought Severity Index. It ranks the magnitude of wetness or dryness over an area on a scale from -6, extremely dry, to +6, extremely wet. Since different parts of the country are on average wetter or drier than others, the Palmer index is calculated so that a given reading relates to the same magnitude of drought anywhere in the nation. *See also: Atmosphere; Blocking Pattern; High Pressure; Low Pressure; Ocean; Precipitation; Rain; Seasons; Snow; Temperature.*

DRYLINE

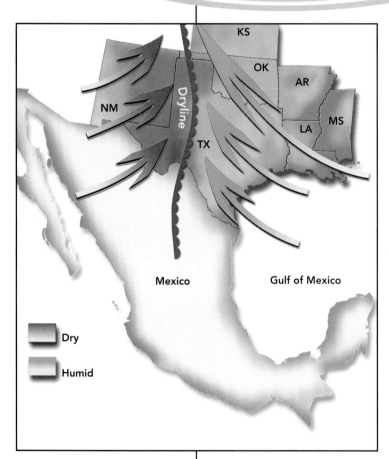

In the United States, drylines usually form in the southern Plains where moist air from the Gulf of Mexico meets up with dry air from the desert Southwest.

A dryline is the atmospheric boundary that forms when moist tropical air meets up with dry continental air. Drylines form in regions of the world where geography allows these two very different kinds of air masses to meet on a regular basis. The dryline is often a focus for the development of severe thunderstorms, which sometimes produce tornadoes, so meteorologists closely monitor its formation and movement.

WHERE DRYLINES ARE FOUND

In the United States, a dryline is commonly found during the spring and summer from western Texas and eastern New Mexico to western Kansas and eastern Colorado. During these seasons, southerly winds from the Gulf of Mexico carry warm, moisture-laden air northward and westward into the southern Plains states. This tropical air mass meets up against very dry air that blows in on southwest winds from Mexico and the Desert Southwest. Where these two air masses meet, the air can change from desertlike to humid and tropical over a horizontal distance of only 50 to 100 miles. The dryline can move back and forth several hundred miles over a period of days, depending on whether the moist or dry flow of air is stronger.

TRACKING DRYLINES

One way meteorologists track the dryline is by analyzing maps that show the atmospheric moisture content over a region. Meteorologists like to use the dew point temperature as a way of measuring the moisture content of air. The dew point is the temperature to which air must cool for the rate of condensation to exceed the rate of evaporation. The lower the dew point, the drier the air. In the spring and summer, dew points over 65°F are considered humid, while those below 40°F are considered dry. On a weather map, meteorologists will watch the southern Plains for a zone where the dew point changes rapidly over a small distance. Often, dew points will fall from the 60s and 70s east of the dryline to 20s and 30s, or even lower, west of the dryline.

DRYLINE AND THUNDERSTORMS

While much of the time the dryline is merely a boundary between dry and humid air, under certain conditions it can serve as the focus for the development of large and powerful thunderstorms. As moist air flows in from the southeast, it collides with the dry air along the dryline and is forced to rise. As the air rises, it expands and cools, and the water vapor inside the air condenses into liquid cloud droplets. If the atmosphere into which the air rises is unstable enough, the rising air can form towering clouds that quickly grow into thunderstorms. An unstable atmosphere is one where temperature decreases rapidly with increasing height; such an environment favors strong upward currents of air. Typically, the atmosphere over a region can become unstable if lower levels of the atmosphere are warmed by the sun from above, while at the same time an area of colder air spreads overhead at high levels of the atmosphere. This kind of situation commonly occurs over the dryline in the southern Plains during the spring and summer. The region of the southern Plains where the dryline most frequently forms is therefore an area where violent thunderstorms commonly occur during the spring and summer. *See also: Air; Air Mass; Atmosphere; Cloud Formation; Dew Point; Seasons; Temperature; Thunderstorms; Water Vapor; Weather Maps.*

DUST STORM

When strong winds blow across a large expanse of dry land, a dust storm can develop. Wind picks up loose dust and sand from the ground and carries it into the air. Wind-blown dust can reduce visibility so severely that vision is restricted to only a few feet. Visibility this low is called zero visibility and poses serious hazards to both automotive and aircraft transportation.

Dust storms are most common in deserts and other arid regions that have an abundance of dust and sand. In the United States, dust storms occur across parts of southern California, Arizona, Nevada, and New Mexico. Elsewhere in the world, they frequently occur in the Sahara region of Africa as well as across the deserts of Australia and Asia. Dust storms are commonly referred to as haboobs, after the Arabic "habub," meaning a violent storm.

How Dust Storms Happen

Dust storms often form in conjunction with thunderstorms. Although thunderstorms are infrequent in desert regions, occasionally the conditions are right for them to form.

First, winds carry moisture-laden air from more humid locations into desert locales. This moisture then heats and rises over the deserts. The water vapor in the air condenses and forms clouds, which quickly develop into thunderstorms. Rain falling out of the base of the thunderstorm drags cooler air down from above. This downrushing air hits the ground and spreads out horizontally, accompanied by strong wind gusts. As this wind spreads out ahead of the storm, it scoops up dust and sand from the ground. If the winds are strong enough, a dust storm can form.

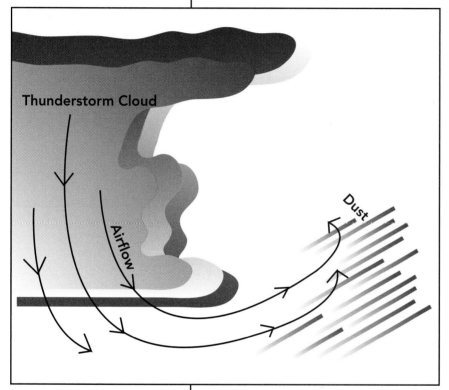

Dust storms often occur in dry areas when air rushing out of a thunderstorm kicks up dirt and dust from the ground.

Wind blowing from thunderstorms is not the only way dust storms can form. Any strong, sustained wind over dry and dusty land can cause a dust storm. Air circulating over hundreds of miles around strong weather systems, such as areas of low or high pressure, can sometimes blow fast enough to pick up dust and sand. Even though the sky may be free of clouds, a dangerous dust storm can develop in these situations if winds are strong enough and the ground is dry enough.

Dust storms that form in this way often appear as an advancing wall of dust thousands of feet high, as the outflow from the thunderstorm plows across the land. These kinds of dust storms can spread out dozens or even hundreds of miles away from the thunderstorm where the outflowing winds originated.

Dangers of Dust Storms

The low visibility can be extremely hazardous to traffic. One of the worst traffic accidents in U.S. history occurred along a stretch of Interstate 5 in central

Dust Storm

California on November 29, 1991. Several years of drought along with freshly plowed fields resulted in a large availability of dust. Winds of 30 to 40 mph blew across the region in the wake of a cold front. Visibility dropped to zero along the highway, and cars and trucks collided with each other in the dust storm. Over 160 vehicles were involved in the collision, which resulted in seventeen deaths and 157

Strong winds blowing across open farmland can lift large amounts of dust into the air.

injuries. *See also: Air; Cloud Formation; Cold Front; Drought; High Pressure; Low Pressure; Rain; Thunderstorms; Water Vapor; Weather; Wind; Visibility.*

EASTERLY WAVE

An easterly wave, also called a tropical wave, is an atmospheric disturbance over tropical regions. Since the air flows from east to west in the tropics, these disturbances move from the east and are therefore called easterly waves. Most easterly waves are relatively weak, causing nothing more than a period of showers and thunderstorms as they pass over a given location. Occasionally, however, an easterly wave can grow into a hurricane as it moves over warm, tropical ocean water. From June to November, meteorologists closely monitor tropical regions for easterly waves and track them once they form.

TRACKING EASTERLY WAVES

There are several things that weather forecasters look for when finding and tracking easterly waves. Satellite photographs are most useful. These photographs, taken from orbiting satellites in outer space, show the detailed patterns of clouds that move around the earth. In the tropics, easterly waves are often marked by clouds, showers, and thunderstorms that are aligned on an axis from south to north, or southwest to northeast. This unsettled weather follows the tropical wave as it moves toward the west through the tropics.

Easterly waves are also associated with a zone of lower atmospheric pressure and winds that veer from northeast (ahead of the easterly wave) to southeast (behind the easterly wave). Meteorologists check weather observations from ships and from stations on land to look for these telltale patterns of wind and pressure. As a tropical wave approaches a location, the atmospheric pressure will fall and winds will blow from the northeast. As the easterly wave passes by, clouds and rain showers will occur as the atmospheric pressure rises and the wind shifts into the southeast.

HOW EASTERLY WAVES FORM

Many of the easterly waves that move across the tropical Atlantic develop as a result of strong temperature differences between the hot Sahara desert in North Africa and the relatively cooler African coastline along the Gulf of Guinea. This temperature difference causes winds to accelerate and come

EASTERLY WAVE

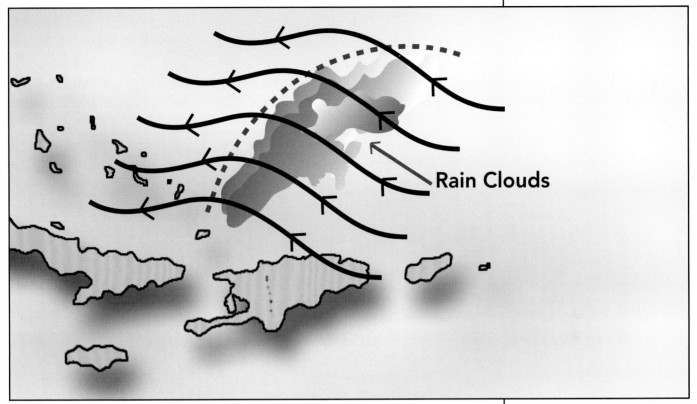

together, or converge, over this region. Converging air rises to form clouds and thunderstorms. From this origin, easterly waves can travel thousands of miles before dissipating.

The vast majority of easterly waves travel harmlessly through the tropics before dissipating. Under certain conditions, however, these disturbances can grow into a hurricane over a period of several days or more. About 60 percent of Atlantic tropical storms and nonmajor hurricanes, with winds of 110 mph or less, originate from easterly waves. About 85 percent of major hurricanes, with winds of more than 110 mph, originate from easterly waves.

Scientists still do not fully understand how these relatively weak weather systems intensify into hurricanes, but there are a couple of necessary conditions. The ocean water over which the easterly wave passes must be warm, with temperatures of 80°F or higher. This warmer water provides heat energy, which assists the development of thunderstorms associated with an easterly wave. Also, winds at high levels of the atmosphere are very important in determining whether an easterly wave grows into a much stronger system. When these upper-level winds are strong and blowing from west to east, they can shear apart developing showers and thunderstorms. On the other hand, a light clockwise circulation of air that spreads apart, or diverges, over an easterly wave removes air from over the developing system. Other air must rise from the surface to take the place of the air that has been removed from over the easterly wave. The water

Easterly waves are disturbances in the flow of air that move from east to west through the tropics.

vapor in this rising air condenses to form showers and thunderstorms. As more air is removed from the top of the system, more air spreads in and rises from below, causing more and more showers and thunderstorms to form. As this process proceeds, an easterly wave can strengthen and eventually develop into a hurricane. *See also: Air; Cloud Formation; Heat Transfer; Hurricane; Ocean; Temperature; Thunderstorms; Satellite; Water Vapor; Weather; Wind.*

EL NIÑO

El Niño refers to an occasional warming of sea surface temperatures in the equatorial Pacific off the northern coast of South America. El Niño, Spanish for "boy child," is a common occurrence each year after Christmas and was named in reference to the Christ child. Sometimes, however, El Niño becomes much stronger and lasts much longer than normal. It is these severe El Niño events that can alter global weather patterns. The major El Niño event of 1997–1998, for example, indirectly caused 2,100 deaths and $33 billion in damage around the world. Scientists are actively studying how El Niño forms and what causes it to occasionally become so severe. Presently there are few answers to these questions.

How El Niño Forms

El Niño is related to patterns of wind and air pressure across the length of the Pacific. Normally, an area of relatively higher air pressure exists over the eastern Pacific in conjunction with colder water off the Pacific coast of South America. At the opposite end of the Pacific, from northern Australia to Indonesia, an area of relatively lower air pressure exists in conjunction with warmer ocean water temperatures. Since air moves from higher to lower pressure, this imbalance in pressure results in winds that blow from east to west across the length of the tropical Pacific. These winds are known as the trade winds.

Off the northern coast of South America, the trade winds blow water at the surface of the ocean westward, away from land. As this water flows slowly toward the west, it gradually warms. In its place, cold water rises from deep in the ocean and flows northward from the southern tip of South America. These currents act to replenish the water that has been spread away from land by the trade winds.

Occasionally, the pattern of air pressure over the Pacific reverses. Pressures rise over northern Australia and southeastern Asia, and fall over the eastern

Pacific off the South American coast. When this happens, the trade winds weaken or even reverse. This allows warmer water from the central and western Pacific to spread eastward into the normally colder parts of the eastern Pacific, resulting in an El Niño event. The cycle whereby air pressure reverses from its normal pattern and then returns back again is called the southern oscillation and can last for several years. Since the southern oscillation and El Niño are so closely linked, scientists refer to them collectively as ENSO events, for El Niño southern oscillation. A major question that remains to be answered is whether the reversal in air pressure and trade winds results in warming of ocean temperatures, or whether warming ocean temperatures causes the reversal of wind and pressure.

EFFECTS OF EL NIÑO

The effects of El Niño are well known and are related to how the ocean interacts with the atmosphere. Warm ocean water raises the temperature of the air in contact with it. Warmer ocean water also contributes more moisture into the atmosphere, since the ocean water molecules move faster on average and more easily evaporate into the air. This added moisture causes an increase in clouds over the eastern equatorial Pacific. As the moisture condenses to form the clouds, it releases heat energy into the air, further warming the atmosphere over the eastern Pacific. The additional heat in the atmosphere increases the contrast in air temperature between equatorial latitudes and middle latitudes over the United States. This difference in temperature strengthens upper-level winds blowing eastward out of the subtropical regions.

Flooding in Central Valley, California, as a result of El Niño.

Storms moving into the West Coast of the United States often draw the abundant tropical El Niño moisture northward, resulting in flooding rains in California during some El Niño years. El Niño-warmed ocean water can also contribute to stronger and more frequent hurricanes in the eastern Pacific. The stronger than normal upper-level winds carry moisture across the southern

United States, where winters on average are much wetter during El Niño events. As these winds blow at high levels over the Atlantic, areas of thunderstorms, which might otherwise develop into hurricanes, are sheared apart and dissipate. Higher than normal pressure over Australia and Indonesia associated with El Niño southern oscillation events suppresses cloud development, so these areas often experience drought conditions during El Niño. Warmer air over Ecuador and Peru aids cloud development, so these areas often experience higher than normal rainfall during El Niño.

Just how El Niño affects other weather patterns around the planet is uncertain. In general, changes in air temperature, pressure, and wind caused by El Niño alter weather patterns over the Pacific. These, in turn, reverberate around the globe. The planetary atmosphere, however, is a complex chain of interaction between air, water, and ocean. It is therefore incorrect to relate El Niño as the sole cause of any specific weather event. Scientists are actively studying El Niño and its causes in order to understand how weather patterns interact around the world. Meteorologists have flown on instrument-laden aircraft into storms fueled by El Niño. They also have deployed floating buoys across the Pacific to monitor ocean temperature and weather conditions. Powerful supercomputers run complex programs that simulate the atmosphere and try to predict how El Niño will affect weather patterns. Discovering how El Niño works will hopefully shed more light on how the earth's oceans interact with its atmosphere. *See also: Air; Atmosphere; Cloud Formation; Drought; Heat Transfer; Oceans; Temperature; Thunderstorms; Trade Winds; Wind.*

EQUINOX

The equinox is a term used to describe a day of the year during which the sun's rays strike the earth directly at a 90° angle along the equator.

There are two equinoxes. In the Northern Hemisphere, the autumnal equinox occurs on September 22 or 23 and marks the first astronomical day of fall. The vernal equinox occurs on March 20 or 21 and marks the first astronomical day of spring.

The equinoxes mark the changing of the seasons in a regular, predictable way. As everyone has experienced, however, weather often does not correspond to the astronomical start or end of a season. Spells of warm weather can occur late into fall, while cold snaps can occur well into spring. Equinoxes and their counterparts, called solstices (the two days of the year, one in summer and one in winter, in which the sun's rays strike the earth directly at their farthest north and farthest south points), are a useful way of understanding how the earth's orbit causes seasons, and therefore weather, to change.

The autumnal equinox signals the beginning of the fall season, colder weather, and colorful leaves.

The angle at which the sun's rays strike the earth is due to the way in which the earth rotates around the sun. The earth is tilted on its axis at an angle of 23.5°. During winter in the Northern Hemisphere, the North Pole points away from the sun. The sun's rays therefore strike the earth most directly in the Southern Hemisphere. During summer in the Northern Hemisphere, the situation is reversed. The North Pole points toward the sun so the sun's rays strike the earth most directly in the Northern Hemisphere. The latitude at which the sun's rays strike the earth directly therefore move from south to north and back again from season to season. The days on which the sun's rays strike the earth directly at the equator are the equinoxes. *See also: Seasons.*

EXTRATROPICAL CYCLONE

Areas of low pressure are called cyclones by meteorologists. Some cyclones, such as hurricanes and typhoons, originate in the tropics and are called tropical cyclones. Other cyclones form farther to the north outside the tropics and are referred to as extratropical cyclones.

Extratropical cyclones carry typical patterns of wind, temperature, and pressure that form and dissipate as the cyclone evolves. In the 1920s, a group of Norwegian meteorologists made the first comprehensive study of the life cycle of extratropical cyclones. Their theory, called polar front theory, is also known as the Norwegian model of extratropical cyclone development. It remains useful today as a general model that describes the evolution of extratropical cyclones from birth to decay.

POLAR FRONT THEORY

Polar front theory is so named because extratropical cyclones form along the polar front. This front is a discontinuous atmospheric boundary that undulates from west to east around the globe at middle and high latitudes. An atmospheric boundary is the transition zone between two masses of air with differing densities. The polar front is a boundary that separates the warmer air that originates in the tropics and subtropics from the colder air that originates near the poles.

In addition to being a boundary between contrasting temperatures, the polar front is a zone of relatively lower air pressure. Extratropical cyclones form along the polar front near the earth's surface in response to wind and temperature patterns at high levels of the atmosphere. Polar front theory describes the stages of extratropical cyclone development and the patterns of wind, temperature, and air pressure that accompany each stage near the earth's surface.

STAGES OF POLAR FRONT THEORY

The first stage of polar front theory occurs before a cyclone develops along the

An extratropical cyclone develops along a front, then dissipates as it becomes detached from the front.

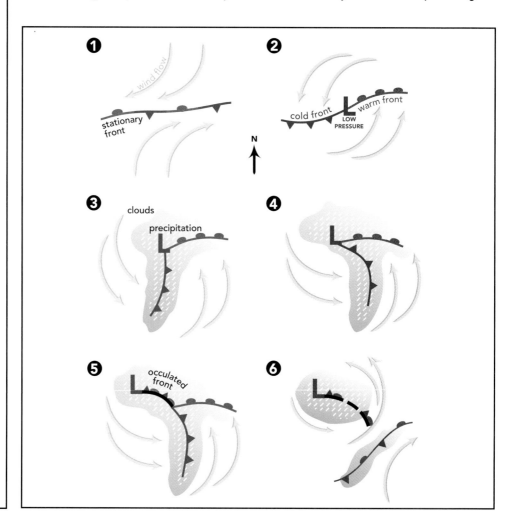

front. North of the front, which lies nearly stationary from west to east, winds blow colder air in from the north and northeast. South of the front, winds blow warmer air from the south and southwest. These opposing wind directions help initiate a rotation in the wind along the front, much like a pencil rotates between a person's hands when the hands move in opposite directions. The direction of this circulation is counterclockwise, which is also called cyclonic circulation by meteorologists.

The second stage of extratropical cyclone development marks the formation of a weak area of low pressure along the front. East of the low-pressure area, the cyclonic circulation causes the polar front to lift northward as warmer air

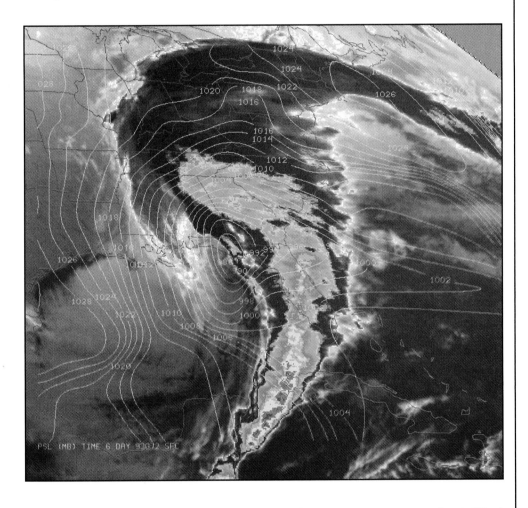

This satellite photograph shows a rapidly strengthening extratropical cyclone across the southeastern United States in March 1993.

moves up from the south. This section of the front becomes a warm front. West of the low-pressure area, the cyclonic circulation causes the front to bow southward as colder air spreads down from the north. This part of the front becomes a cold front.

Next, the extratropical cyclone becomes stronger as pressures fall. Meteorologists refer to this as a "deepening" process. Air continues to spiral cyclonically around the storm and in toward its center. Warm, moist air spreading northward ahead of the low-pressure area gradually rides over the denser, colder air along the warm front. As the air rises, it cools, and the moisture in the air condenses into clouds that produce steady rain or snow. Along the warm front, the warm, less dense air moving up from the south gradually spreads northward over the colder air. Therefore, the warm front slowly moves north. Along the cold front, by contrast, denser, colder air spreading in from the north and west acts like a plow as it wedges underneath the warmer air. Air is consequently forced to rise faster along the cold front, so precipitation is usually heavier and more showery than along the warm front. In between the warm and cold fronts, the zone of warm air is known as the warm sector of the extratropical cyclone.

As the cyclone matures, the air pressure in the center of the low-pressure area drops even farther. The difference between the low air pressure in the center of the storm and the higher air pressure surrounding the storm increases. Since air moves from high to low pressure, wind circulates faster and faster in toward the center of the storm. Precipitation becomes heavier and more widespread along the warm and cold fronts as more and more air is forced to rise. At its strongest, an extratropical cyclone and its accompanying fronts can spread rain or snow over thousands of miles.

The extratropical cyclone begins to weaken as soon as the fast-moving cold front overtakes the slower warm front. This first happens just south of the area of low pressure, then farther and farther south along the fronts. Where the cold and warm fronts meet, a new kind of front forms called an occluded front. The occluded front becomes longer and longer as the cold front continues to catch up to the warm front. This causes the area of low pressure to become farther and farther removed from the warm sector. Since the warm and moist air serves as an energy source for the cyclone, it begins to weaken, or fill, as it occludes and drifts north.

Finally, the weakening cyclone becomes completely detached from the front as the circulation around the cyclone thoroughly mixes the warm and cold air. It drifts northward and dissipates, while the polar front settles to the south. In another couple of days, a new area of low pressure may form along the front.

VARIATIONS ON THE THEORY

The polar front theory is a simplified progression of events in the life cycle of an extratropical cyclone. Depending on where the cyclones form, the patterns of wind, temperature, and fronts can be very different from this model. For example, cyclones moving toward the eastern United States from the Plains states are known to dissipate upon reaching the Appalachians, then appear to "jump" the

mountains and transfer their energy to a new cyclone that can develop rapidly along the East Coast. This process is extremely complex and not fully understood. Polar front theory, which was developed using observations over a limited section of the earth, fails to describe variations in extratropical cyclone development that occur in different parts of the world. As a general model of how cyclones develop, however, it remains useful decades after it was conceived.
See also: Air; Atmosphere; Cold Front; Cloud Formation; Low Pressure; Occluded Front; Precipitation; Rain; Snow; Temperature; Warm Front; Wind.

FLOOD

A flood occurs when a body of water overflows its natural boundaries. Flooding is the leading cause of weather-related fatalities in the United States, with an average of 140 deaths annually. By comparison, tornadoes kill an average of seventy people each year, and hurricanes cause an average of around twenty-five deaths each year. There are many reasons why floods occur, but excessive rain or snow are typical factors. Floods are classified by their cause and by the type of body of water from which they occur.

River floods occur along larger rivers, usually after a prolonged period of above-normal precipitation. Large and long rivers are normally accustomed to carrying such an enormous quantity of water that it takes weeks or months of above-normal rainfall to cause a flood. Heavy rain does not need to fall on the river itself for flooding to occur; it can fall on the vast expanse of land that drains into the river. Rapid melting of deep snow can also lead to river flooding.

Flash floods occur along smaller rivers, streams, and creeks. As the name implies, water levels rise very rapidly and often unexpectedly. These kinds of floods usually occur when strong and slow-moving thunderstorms quickly dump large amounts of rain on a stream or small river. They can also be caused by dam failures. Flash floods do not affect nearly as large an area as a river flood, but they are usually more deadly because they strike suddenly and unexpectedly.

Smaller in scope but still potentially deadly, urban flooding occurs when a city's drainage system becomes overwhelmed or clogged in very heavy rainfall. Water then backs up and floods low-lying areas, such as dips in roads underneath bridges. Many flood-related fatalities have occurred when unsuspecting motorists drive into flooded roads not knowing how deep the water is.

Coastal floods occur when strong winds pile up ocean water along the shore.

Extreme flooding can even wash away homes.

This situation often happens when strong areas of low pressure move along the coast, or when hurricanes or tropical storms cross the coastline. Coastal flooding is at its worst during high tide.

Floods in Recent U.S. History

Floods have affected all regions of the country. Although flooding can occur at any time of year, certain kinds of floods are more likely to happen during certain seasons. Flash floods, for instance, are usually caused by torrential thunderstorm rainfall. Since thunderstorms are primarily a warm-season phenomenon, flash floods usually happen during spring and summer. River floods, on the other hand, can happen during any season since they are caused mainly by prolonged, heavy precipitation. Coastal flooding occurs most frequently when strong winds blow onshore along low-lying coastal areas. This most often occurs with coastal storms that form from the fall through the winter. In the United States there have been thousands of flooding events over the years, some of which have achieved historical status because of their magnitude and resulting damage or loss of life.

The most extreme example of a river flood in recent times was the Mississippi River flood in the summer of 1993. During this event, the Mississippi and other major rivers overflowed their banks and set new flood level records all across the Midwest. The flooding lasted for weeks as whole towns along the river were inundated. Over 17 million acres of land were flooded by rivers across nine states, and over 22,000 homes were damaged or destroyed. The Missouri River, for instance, normally no wider than a half of a mile, expanded to as much as five to ten miles wide in places. This 1993 river flooding was the result of

Buildings along the banks of rivers are particularly vulnerable to flooding.

heavy thunderstorms that persisted over the Midwest from the late spring through the summer. While thunderstorms normally occur from time to time across the Midwest in spring and summer, in 1993 they developed over the same areas day after day. Each round of thunderstorms dropped torrential rain that drained into the same rivers, causing them to rise higher and higher. The effects of the above-normal rainfall were worsened since the ground was already saturated from heavy precipitation as far back as the previous autumn.

Another example of river flooding occurred in spring 1997 along the Red River in parts of North Dakota, Minnesota, and Canada. This flood was caused by extremely heavy snowfall across the northern Plains through the winter months. The snow contained an enormous amount of water, which, when the snow melted, was released into streams and rivers across the region. To make matters worse, another snowstorm dumped up to three additional feet of snow across the region in April. When all the snow finally melted, disastrous river flooding occurred. The Red River reached a height of well over twenty feet above normal, flooding towns, cities, and farmland. Grand Forks, North Dakota, was hit particularly hard, with over 90 percent of the city below water at one time.

An example of a disastrous flash flood occurred on July 31, 1976. As much as twelve inches of rain fell in four hours on Big Thompson River in Colorado from a thunderstorm downpour. This water raced down the mountainsides and into the narrow river canyon, resulting in a wall of water that swept downstream. Houses, roads, and cars were swept away, and 135 people lost their lives.

Coastal flooding can occur not only with hurricanes but with any strong storm system that blows ocean water into coastal regions. In March 1993, for example, an unusually strong area of low pressure developed rapidly across the northern Gulf of Mexico. This storm, which would go on to become known as the Storm of the Century across many parts of the eastern United States, did its first major damage in Florida. Strong southerly winds ahead of the storm across the eastern Gulf of Mexico drove water northward along Florida's western coastline. Then, as a powerful cold front swept across the area, gale force winds out of the west pushed the water inland. Ocean water rose as much as twelve feet above normal in some areas, inundating houses and roads. Seven people were killed in this unexpected surge of water.

FORECASTING FLOODS

There are a number of techniques that meteorologists use to predict flooding in a given area. They study radar and satellite images, observations of current weather conditions, and output from supercomputers that predict weather patterns across the country.

Radar is a helpful tool for predicting flooding conditions over short time spans. It is particularly useful for anticipating the flash flooding that can occur when extremely heavy rain falls in a short time. Radar detects rainfall by bouncing pulses of microwave energy off falling raindrops or snowflakes. Some of the energy bounces back to the radar dish where it is received and analyzed. Computers convert the signals reflected from the precipitation into a graphical display that shows the location and intensity of the precipitation. In spring and summer, meteorologists watch time-lapse loops of radar data to track the movement of thunderstorms. It is the heavy rainfall from these storms that most often produces flash flooding, especially if the storm moves very slowly or stalls over an area.

Meteorologists also study satellite photographs that display the infrared radiation energy emitted by clouds and by moisture in the atmosphere. These images allow meteorologists to view cloud patterns that would otherwise

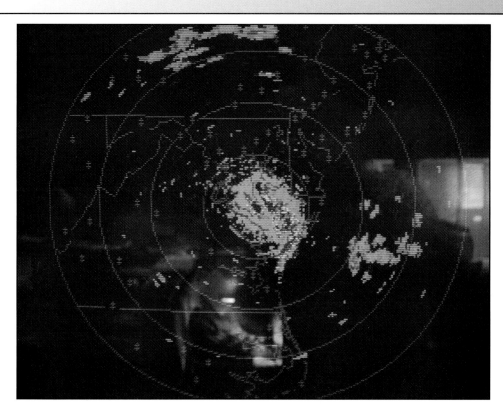

Radars help meteorologists track heavy rain which may lead to flooding.

become invisible in the dark of night. Infrared, or IR, satellite images show the temperature of cloud tops. Clouds that produce particularly heavy rain often rise to great heights in the atmosphere and therefore become very cold. IR images can also detect not only clouds but the normally invisible water vapor molecules in the atmosphere. Even though the sky may be free of clouds over an area, an IR water vapor image can show that large quantities of gaseous water vapor are collecting over a region. This water vapor may contribute to heavy rainfall if thunderstorms form in the area.

Large supercomputers perform extremely complex mathematical equations to predict weather conditions over large parts of the globe, not only at the surface of the earth but also high in the atmosphere. Meteorologists examine the output of these computer models to see where the computer predicts heavy rainfall. They check to see if the results from the computer programs correspond to actual conditions that are being reported at thousands of weather observing sites. These sites report not only temperature, wind, air pressure, and moisture levels but also the amount of rain that has fallen.

In addition to meteorologists, hydrologists—scientists who study the natural flow of water—collect data that can help predict floods. Hydrologists take measurements of soil moisture, river levels, and snow depth to asses how water will behave as it drains from land areas into streams and rivers. Soil moisture is important because heavy rain that falls on water-logged ground will lead to flooding much more quickly. Hydrologists also constantly monitor water levels along streams and rivers. They know the exact water levels at which streams and rivers in their forecasting area begin to flood, and how much rainfall is needed to produce flooding in a given area. Since runoff from melting snow flows into streams and rivers, hydrologists also monitor snowfall levels. On average, ten inches of snow will melt into one inch of water, and this meltwater can contribute to flooding when deep snowfall melts rapidly.

The National Weather Service issues a variety of advisories to warn the public of potential or imminent flooding. A flood or flash-flood watch means that atmospheric conditions are favorable for heavy rain that may lead to flooding. A flood or flash-flood warning means that flooding is occurring or is imminent. An urban and small stream flood advisory warns of possible flooding due to poor drainage in cities. *See also: Cold Front; Low Pressure; National Weather Service; Oceans; Precipitation; Radar; Rain; Satellite; Snow; Thunderstorms; Tornadoes; Water Vapor.*

FOG

Fog consists of billions of tiny water droplets suspended in air. Essentially, fog is a cloud at the earth's surface. Fog can significantly reduce visibility, making driving extremely hazardous. Dense fog can also disrupt air and ship transportation.

How Fog Forms

Fog forms when invisible water vapor molecules condense into liquid water droplets. Condensation is the process whereby gaseous water vapor changes into liquid water. This happens when water vapor molecules adhere to one another, or to other objects, when they collide. The slower the water vapor molecules move, the more likely they are to stick together upon colliding with one another or with other particles in the air. Air molecules move more slowly as the temperature falls, so the condensation that produces fog typically occurs during the coolest parts of the day—the late night and early morning.

A clear night with light winds and moist ground is most favorable for the development of fog. A mostly clear nighttime sky allows the ground to radiate its heat energy into space. As the ground becomes cooler, so does the air in contact with it. (Clouds, on the other hand, radiate their own energy back to the earth, causing the ground to remain warmer than it would otherwise be.) The cooler the air becomes, the more likely it is that the water vapor in the air will condense into fog.

Fog obscures the Golden Gate Bridge in downtown San Francisco.

Recent rains and light winds assist in the formation of fog. As the slight wind brings air in contact with the moist ground, it picks up more moisture and increases the humidity of the air. Fog that forms as a result of this radiational cooling of the ground is called radiation fog.

OTHER KINDS OF FOG

Air can also cool if it moves over an already cold body of land or water. Fog that forms in this way is called advection fog. This kind of fog frequently occurs along the Pacific coast, as warmer air spreads over much colder ocean water. As the air cools, the water vapor in the air condenses into fog. Advection fog also occurs in winter when warm, moist air from the south spreads over snow cover.

If moist air flows uphill, it can form what is known as upslope fog. As air moves uphill it moves into lower atmospheric pressure, since less of the atmosphere weighs down from overhead. The air expands as it moves into lower pressure, and air molecules become farther apart and move more slowly. The air therefore cools, condensation occurs, and fog forms. Areas along the slopes of mountains can experience this kind of fog if moist winds blow uphill.

Mixing fog occurs when warm, moist air mixes with colder air. A small-scale example of mixing fog is a person's breath seen on a cold day. Mixing fog often occurs in late fall and early winter as cold air moves over relatively warmer, unfrozen lakes. Steam will rise from the lakes as the cold air mixes with water vapor from the lake below.

The West Coast is the foggiest place in the United States because of the frequency of advection fog. Other foggy areas include the Appalachians and New England. These regions have numerous valleys into which cold air drains, contributing to the frequent formation of radiation fog. *See also: Advection; Atmosphere; Heat Transfer; Oceans; Radiation; Snow; Visibility.*

FRANKLIN, BENJAMIN

Benjamin Franklin (1706–1790) was an early American weather enthusiast.

Benjamin Franklin (1706–1790) was an American inventor, statesman, scientist, and writer. In addition to his many other accomplishments, he was the foremost weather researcher and observer in colonial times.

Franklin is perhaps best known for his experiments with lightning. In June of 1752, Franklin attempted to prove his hypothesis that lightning is electrical in nature. He flew a kite in a thunderstorm near his home in southeastern Pennsylvania, hoping that lightning would strike the kite. As the kite flew high in the sky near the clouds, Franklin noticed loose threads of the kite standing on end. As he touched his hand to the end of the string, he felt a slight shock. While Franklin was able to prove that lightning is indeed an electrical phenomenon, we know today that his experiment was extremely dangerous. In fact, a short time after Franklin flew his kite near a thunderstorm in America, a scientist in Russia attempted to re-create Franklin's experiment and was killed. Benjamin Franklin, on the other hand, went on to invent the lightning rod as a form of lightning protection that could be used on houses. A lightning rod is a metal pole attached to the top of a house, usually at the chimney, and extended down into the ground. If lightning strikes the house, it is attracted to the metal rod, which safely channels the electrical energy into the ground.

Franklin also was the first to detect the Gulf Stream, a warm current of water that flows northward off the East Coast of the United States. He did this by taking temperature readings of the ocean water while on a ship at sea. Another meteorological discovery credited to Benjamin Franklin was the observation that storms, and weather in general, move from west to east. In the eighteenth century and earlier, this was not immediately apparent since a reliable network of weather observers had not been created. Using his discovery of the movement of weather systems, Franklin was able to make some of the first recorded weather forecasts in America.

Franklin's curiosity about the weather led him to keep a weather journal for many years. A weather journal is simply a written account of the day's weather at the observer's location. Weather historians have found Franklin's records invaluable as a source of information on the climate of the eastern United States during colonial times. *See also: Climate; Lightning; Oceans; Temperature; Thunderstorm; Weather.*

FREEZING RAIN

Freezing rain is a form of precipitation that occurs during colder months. It consists of falling water droplets that freeze on impact with the ground. Heavy freezing rain is known as an ice storm, with trees and power lines suffering damage as the result of being coated with a thick layer of ice.

Freezing rain forms when snow falls through an inversion. An inversion is a layer of air, often hundreds or thousands of feet thick, in which air temperature rises with increasing altitude. Normally, air temperature decreases with increasing altitude. If the inversion is thick enough, the snow falling through the inversion completely melts. Studies have shown that at least 300 meters of above-freezing air is required for falling snow to melt into rain.

For freezing rain to occur, the melted snow needs to fall into a shallow, colder layer of air beneath the inversion. (If this layer of subfreezing air is too deep, the melted snow will refreeze into ice pellets, or sleet—it has to be be shallow

Freezing rain forms when snow melts into rain as it falls through an inversion, then refreezes when it hits the ground.

One example of a particularly damaging ice storm occurred in early January 1998. Unseasonably warm, moist air flowing into the northeastern United States on southerly winds spread over low-level arctic air moving south from Canada. The result was a heavy and prolonged period of freezing rain. The total damage was well over $2 billion.

enough that the water drops do not have time to refreeze before hitting the ground.) Once the water drops do make contact with the cold ground, they refreeze into a layer of ice on any exposed surface.

The conditions that support freezing rain occur in winter when relatively mild air spreads northward over dense, cold air near the earth's surface. Sometimes local geography aids the formation of freezing rain, as when mild air overrides cold, dense air trapped in valleys. Freezing rain often occurs just north of an advancing warm front. In these situations, warmer air moving northward rides up and over the cold surface air north of the warm front. Snow melts as it falls through this warmer layer of air north of the front, then refreezes as it hits the surface. If the warm front is moving very slowly, an extended period of freezing rain can occur. *See also: Air; Precipitation; Rain; Sleet; Snow; Temperature; Warm Front.*

FROST

Frost consists of tiny ice crystals that form on exposed outdoor surfaces. Frost usually forms during colder months as overnight temperatures drop near or below freezing. It may damage or kill plants that grow close to the ground, such as garden vegetables.

Both frost and dew form after sunset, as the temperature of the ground falls. As the ground temperature drops, so does that of the air in contact with it. Water vapor molecules in the cooling air move at slower and slower speeds, becoming more likely to stick to each other or other objects when they collide. The temperature at which the water vapor molecules move slowly enough that they adhere more often than they fly apart is known as the dew point. The frost point is the temperature at which the water vapor in the air changes directly into ice crystals in a process called deposition. These ice crystals can form on grass, cars, and other surfaces.

Certain conditions favor the formation of frost. A nighttime sky that is cloud-free will allow greater surface cooling. Clouds, like all objects, radiate a certain amount of energy. When clouds are present, the ground absorbs some of the energy and remains relatively warmer. The warmer the ground, the less likely the air in contact with it will be able to cool toward the frost point.

A calm night also aids in frost formation. Wind mixes cold air near the surface of the earth with relatively warmer air just above it. This keeps the air temperature slightly higher than when calm conditions allow cold air to settle near the ground.

FROST

Frost can form when the average air temperature remains a few degrees above freezing. In these situations, the temperature of the air at eye-level might be 35°F. Near the ground, however, a layer of subfreezing air only a few feet thick allows frost to form.

Sometimes the temperature falls below freezing but does not reach the frost point. This can happen when cold air is particularly dry. With relatively scant

Frost consists of tiny, delicate ice crystals.

water vapor molecules in the air, not enough moisture is present to form frost. This condition is called a freeze and can be very dangerous to crops. Growers of fruit trees in Florida and California use heaters in their fields when a freeze is imminent. Sometimes they spray water on the trees to form a protective coating of ice on their fruit. The ice ensures that the temperature of the fruit remains near freezing, while the outside air becomes much colder. *See also: Air; Dew; Dew Point; Frost; Frostbite; Temperature; Water Vapor; Wind.*

FROSTBITE

Frostbite occurs when skin and underlying body tissue freezes as a result of exposure to extremely cold air. It is a serious health hazard in winter during cold waves when temperatures fall well below normal. While cold air alone can lead to frostbite, the combination of cold and wind is more dangerous since wind strips heat from the body. Frostbite most commonly affects the extremities of the body, such as toes, fingers, ears, and the nose, since these areas are most exposed to cold air.

There are several symptoms of frostbite. As skin and body tissue is exposed to very cold air, its temperature drops. Blood flow slows, then stops. This causes skin and body tissue to die. The first symptom of frostbite is tingling skin. This can become painful and eventually lead to a loss of feeling. Skin becomes whitish and swollen, then turns hard, pale, and cold. Blistering may occur.

Frostbite can be treated successfully if several steps are followed. If the frostbitten person is far from medical help, the body should not be quickly thawed or exposed to direct heat, such as a heater or furnace. Nothing should be rubbed on the affected skin. If outdoors, use a warmer part of the body to heat the frostbitten area. Place a frozen hand underneath the armpit, for example, or between the thighs. If indoors, immerse the affected area in warm—but not hot—water, or cover with warm blankets. Have the person drink warm, but again not hot, fluids. It may take twenty to forty-five minutes for the frozen area to thaw. Do not break any blisters, and do not place any bandages or ointments on the frostbitten skin. Medical help should be sought as soon as possible. *See also: Air; Cold Waves; Temperature.*

FUJITA, THEODORE

Theodore Fujita (1920–1998) was a professor of meteorology at the University of Chicago and pioneer in the research of severe local storms. He made important discoveries about thunderstorms, tornadoes, and downbursts.

While in his native country of Japan, Fujita made observations of thunderstorms and wrote two research papers on the subject. He sent them to Dr. Horace Byers, the world's foremost expert on thunderstorms at that time, at the University of Chicago. Eventually Byers arranged for Fujita to study at the University of Chicago, where he ultimately became professor of meteorology and a U.S. citizen.

Fujita's early work centered on the analysis of thunderstorms and how they alter the air pressure and winds over a relatively small surrounding area. While large weather systems can be tracked across continents, small weather events such as thunderstorms take place on scales of tens of miles. This scale of observation is called mesoscale, and Fujita was one of the first to make detailed mesoscale analyses of thunderstorms.

Fujita was also very interested in tornadoes, the study of which became a major part of his life's work. From the 1960s through 1990s, Fujita mapped the tracks of hundreds of tornadoes. He flew over the tornado paths in a Cessna aircraft only days after the tornadoes had touched down, taking photographs and plotting the storms' tracks on maps. Fujita mapped the century's largest tornado outbreak, occurring April 3–4, 1974. Thirteen states and Canada were struck by 148 tornadoes, killing over 300 people.

While studying tornadoes by air, Fujita noticed unusual swaths of damage left on the ground. These swaths were visible inside the path of the tornado and sometimes appeared as looping streaks similar in appearance to a looping telephone cord. Fujita hypothesized that these streaks were made by smaller "minitornadoes" that spun inside the larger tornado funnel. He called these minitornadoes "suction vortices," and named the peculiar swaths of damage they produced "suction swaths." Through the 1970s and 1980s, photographs and videos of tornadoes provided evidence that proved Fujita's hypothesis. These vortices are very small but intense—they can destroy a house while leaving the house next door suffering only light damage. Fujita's study of a tornado in Lubbock, Texas, in 1970 demonstrated that most of the fatalities from the tornado occurred inside suction swaths.

Another of Fujita's major discoveries concerned a different thunderstorm-related phenomenon. In June 1975, an Eastern Airlines jet crashed while

Tornado damage is ranked on the Fujita Scale, or F-Scale, devised by Theodore Fujita.

attempting to land at John F. Kennedy airport in New York City, killing 122 people. Thunderstorms had been in the area while the plane was making its final approach, so Fujita investigated the crash from a meteorological perspective. He speculated that the plane had flown into a strong, outflow of air plunging down from the base of the thunderstorm. Fujita termed this phenomenon a downburst and published his findings. While the scientific community was initially skeptical, subsequent jet crashes, including one in Dallas in 1985, proved the existence of downbursts and the extreme hazard they pose to aircraft. Because of Fujita's findings, special wind-sensing instruments have been installed at all major airports. These devices can detect the pattern of wind associated with downbursts and provide early warning when downbursts occur. Pilots have also been trained in the dangers of downburst winds and know what to do if they are caught in one.

See also: Downburst; Meteorology; Thunderstorms; Tornadoes.

General Circulation

The flow of air around the earth averaged over a long period of time is called the general circulation. Winds vary from day to day on a local level, depending on the movement of weather systems. When these variations are averaged out over many years, a global circulation pattern emerges. In this general circulation, air rising in one part of the globe compensates for air sinking in another part of the globe. At the same time, eastward moving air at one latitude compensates for westward moving air at another. The general circulation helps explain why some parts of the earth are normally dry while others are wet.

The Path of General Circulation

While there is no beginning or ending point in the global circulation of air, one place to start when describing the general circulation is the tropics. Air warmed by the sun near the equator tends to expand and rise as the air molecules move faster and the air becomes less dense. Thus, in the tropics there exists a zone of rising air and low pressure, accompanied by clouds, showers, and thunderstorms. To replace this rising air, other air flows in horizontally near the earth's surface from the northeast and southeast. These winds are known as the trade winds. The zone of rising air and clouds into which the trade winds flow is known as the intertropical convergence zone, or ITCZ.

The general circulation of the atmosphere contains distinct zones where air rises, sinks, and blows from certain directions.

The air that rises along the ITCZ eventually hits a layer of warmer air at a high level of the atmosphere called the stratosphere. The stratosphere is considered a stable layer of air because it prevents air from moving upward or downward. Air rising from below hits the stratosphere and is forced to spread out horizontally toward the north and the south. As this horizontally spreading air flows hundreds of miles toward the poles, it follows the lines of longitude to converge at the poles. The poleward moving air also becomes cooler. As the air converges and cools, it increases in density, causing it to sink back down toward the surface of the earth. This sinking air occurs in the subtropics near 30° north and south latitude. As the sinking air hits the earth's surface, it spreads apart horizontally. Some of the air moves southward, eventually flowing into the tropics and rising again along the ITCZ.

Some of the air that sinks in the subtropics hits the earth and spreads northeastward. Meanwhile, near the poles, dense cold air sinks and spreads southeast. This southward-moving air collides with the northward-moving air along a boundary called the polar front. The position of this front varies by season and it undulates northward and southward as weather systems move across the middle latitudes. As air collides along the polar front, it rises, cools, and expands once again. The water vapor in rising, cooling air condenses into rain-producing clouds. As in the tropics, this rising air spreads apart horizontally upon hitting the stratosphere. Some of the air spreads southward and some spreads northward, eventually sinking at the poles.

General Circulation and Weather

The global circulation of air directly influences the climate across the globe. Along the intertropical convergence zone, rising air forms showers and thunderstorms, which contribute to a wet climate in many parts of the tropics. This part of the world, therefore, contains most of the world's rainforests. Near 30° north and south latitude, by contrast, sinking air prevents clouds from forming. Most of the world's major deserts are located in this zone. ***See also: Air; Atmosphere; Cloud Formation; Intertropical Convergence Zone; Low Pressure; Thunderstorms; Trade Winds; Water Vapor.***

Global Warming

Scientists have observed a gradual increase in the average temperature of the earth during the twentieth century. This trend is referred to as global warming, and it is predicted to continue through the twenty-first century. Questions exist as to how much of the observed global warming has been due to increased pollution, and how much is due to natural causes. Some of the destructive consequences of global warming include an increase in ocean water levels, shifting climatic zones, and more vigorous storms. Scientists who study weather and climate are as yet unsure what the magnitude of these effects will be.

Global Warming

Scientists are sure, however, that temperatures are rising. The global mean surface temperature has increased 0.6° to 1.2°F since the late nineteenth century, a time coincident with the start of the industrial revolution. About 0.5°F of this warming has occurred during the last forty years, when scientists started making reliable global temperature calculations. The ten warmest years of the twentieth century, averaged for the planet as a whole, have all occurred since 1980. Scientists have also observed a slight increase in cloud cover (indicating that warmer temperatures are evaporating more water) and a slight decrease in snow cover across the planet since 1987. Although the earth's average temperature is warming, local and regional variations in weather patterns can still lead to severe winters and snowstorms.

Causes of Global Warming

There are many explanations for the increase in temperature, but scientists are unsure as to the relative importance of each cause. One possible cause is the increase in the concentration of certain atmospheric gases. These gases, called greenhouse gases, are effective absorbers of outgoing radiation from the earth.

One of the gases thought to contribute to global warming is carbon dioxide. Evidence for an increase in carbon dioxide in the atmosphere comes from a couple of sources. One is the Mauna Loa Observatory on Hawaii. This research station is located on a volcanic mountain in the middle of the Pacific Ocean, allowing scientists to sample air with a purity that is difficult to obtain from areas much closer to cities and industry. Observations since 1958 have shown an obvious and significant increase in carbon dioxide levels.

Scientists learn about levels of atmospheric carbon dioxide from years past by examining ice from Arctic regions. Deep holes are drilled into ice sheets, and an ice core is removed and studied. The scientists probe tiny air bubbles trapped inside the ice for carbon dioxide concentrations. The deeper in the ice these bubbles are, the farther back in time the air was trapped from the atmosphere. Studying ice cores in this way, scientists have measured an increase in atmospheric carbon dioxide levels since the late nineteenth century.

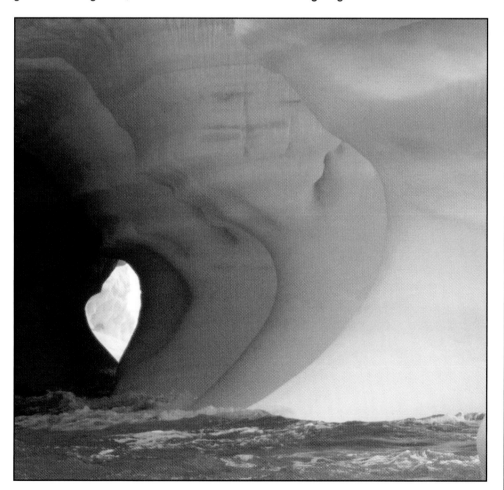

Global warming may lead to an increase in ocean water levels as glaciers and ice sheets melt.

The results of human activity, such as the industrial and automotive combustion of fossil fuels, may be increasing the concentrations of certain greenhouse gases. The gases include carbon dioxide, methane, nitrous oxide, and chlorofluorocarbons (CFCs). The higher the concentration of greenhouse gases in the atmosphere, the more outgoing energy is absorbed. The absorbed energy causes air molecules to move around faster, resulting in a slightly higher air temperature.

The atmosphere may also be warming because of natural causes. Most scientists agree that the planet is still emerging from the most recent ice age, which peaked about 18,000 years ago. Periods of colder and warmer global weather occur naturally over the billions of years of the planet's history. They occur slowly, over thousands or even millions of years. One reason for these temperature swings is subtle changes in the earth's orbit around the sun. The path that the earth takes, the angle at which it is tilted, and the variation in this tilt all influence the amount of solar energy that reaches the earth. Significant changes in the amount of solar energy received by the Earth can substantially alter the temperature of the earth.

Effects of Global Warming

Continued global warming is widely believed to have destructive consequences. One result of global warming is the melting of polar ice caps. This would increase ocean levels, perhaps to the point where low-lying coastal regions will become flooded. Global warming will also have an impact on agriculture. Regions of widespread farming may become too hot and dry to support crops if global temperatures rise. Also, higher global temperatures may fuel stronger storms, like hurricanes or winter storms. Major questions exist as to how much warming is needed to cause these effects and whether our current rate of global warming will slow, increase, or remain the same.

The way in which the atmosphere interacts with the oceans poses some questions as to the effects of global warming. A warmer ocean, for instance, will result in more ocean water evaporating into the atmosphere. This additional moisture in the air may result in increased cloudiness, which could block sunshine and in turn limit the amount of global warming that takes place.

Meteorologists attempt to predict the consequences of global warming by using computers to run complex software that simulates the behavior of the atmosphere under the influence of increased greenhouse gases. Politicians, meanwhile, are attempting to create policies and laws that will limit the increase in greenhouse gases. ***See also: Atmosphere; Climate; Greenhouse Effect; Ice Age; Oceans; Pollution; Radiation; Temperature; Weather; Winter Storms.***

GRAUPEL

Graupel is a frozen form of precipitation that appears as a small icy or snowlike pellet. While it usually does not fall to the earth in large quantities, it plays a role in the formation of rain, snow, and hail inside clouds. It also may contribute to the electrification of clouds.

Graupel is thought to play a role in the formation of lightning.

Graupel forms when small ice crystals or snowflakes inside a cloud fall into an area of supercooled water droplets. Supercooled droplets are tiny drops of water that remain in a liquid state well below freezing. For a droplet to freeze into ice, a certain amount of water molecules must lock into a rigid crystalline structure. In extremely small droplets, fewer molecules are available to do this. The constant jostling of molecules in a droplet ensures that freezing does not occur even at temperatures as low as -30°F.

When an ice or snow crystal falls into a region of supercooled droplets, the droplets collide with the crystal and freeze on contact. This process is known as riming. Graupel forms when enough riming occurs that the original ice or snow crystal is completely coated with tiny droplets of ice.

While graupel can fall to the earth as precipitation, it plays a more important role in the processes that occur inside clouds. Graupel often breaks apart inside the cloud after significant riming has occurred. The splinters of ice that break away cause other supercooled droplets to freeze. This process can happen over and over again, forming more ice crystals in a cloud than there would be in the absence of graupel. Ice crystals can then grow larger and form into snowflakes by attracting water vapor molecules away from supercooled droplets. The ice can fall to the ground as snow or melt as it falls and become rain.

Graupel can also grow into hail if it is caught in the strong upward currents of air inside a thunderstorm. As the graupel is born aloft inside the thunderstorm updraft, large amounts of supercooled water collide with the graupel. The graupel is coated with an increasing layer of ice and becomes a hailstone. Eventually, the vertical air currents cannot support the weight of the hailstones and they fall to earth.

In the right conditions, graupel can become hail.

The ability of thunderstorms to produce lightning may be due, at least in part, to graupel. As supercooled water droplets collide with graupel and freeze, they release a small amount of heat. This causes the surface of graupel to become relatively warmer than other, smaller ice crystals inside the cloud. As warmer graupel passes near or collides with colder ice crystals, positive ions move from the warmer graupel to the colder ice crystal. This causes graupel to become negatively charged, and ice crystals positively charged. Since the ice crystals are lighter than the graupel, vertical air currents carry them aloft into the higher parts of the thunderstorm clouds. The heavier graupel remains in the middle and lower parts of the cloud. This sets up a difference in charge between upper and lower parts of a thunderstorm cloud that can lead to the generation of lightning. *See also: Air; Hail; Lightning; Precipitation; Rain; Snow; Thunderstorms; Water Vapor.*

GREENHOUSE EFFECT

The greenhouse effect is the result of a process by which certain atmospheric gases warm the planet. The greenhouse effect is a natural process that helps regulate the earth's temperature and keeps the earth warm enough for life to exist. Concern over the greenhouse effect has arisen because humans have increased the concentrations of some of the gases responsible for the greenhouse effect. Scientists hypothesize that higher concentrations of these gases increase the greenhouse effect and will result in a warmer planet. There is much speculation and uncertainty about the possible destructive effects of an increased greenhouse effect.

The greenhouse effect occurs as a result of the exchange of radiation energy between the sun, the earth, and the earth's atmosphere. Like all things, the sun, the earth, trees, and people emit a certain amount of radiation energy. The sun emits radiation energy that, for the most part, passes through the atmosphere and is absorbed by the earth. The atmosphere also emits radiation energy that is absorbed by the earth. In fact, because the sun covers a much smaller fraction of the sky than the atmosphere, the earth receives almost twice the amount of radiation energy from the atmosphere than it does from the sun.

The radiation energy that the earth receives from the sun and the atmosphere warms the ground and the oceans. The warmed ground and oceans, in turn, warm the layer of the atmosphere just above them. This heat in the lower part of the atmosphere is distributed to higher parts of the atmosphere by vertical air currents. Therefore, the atmosphere is not heated directly by sunshine but by the warm ground and oceans underneath it.

Since the ground gains twice as much energy from the atmosphere as it does from the sun, any change in the amount of energy emitted by the atmosphere will significantly warm the earth and the air that is in contact with it. The atmosphere can emit more energy if it is itself more energetic. This can happen if the atmosphere absorbs more energy from other sources.

This is the role that greenhouse gases play in the atmosphere: They absorb the energy that radiates from the earth. The gases that are most effective in absorbing this outgoing energy are water vapor, carbon dioxide, methane, nitrous oxide, and chlorofluorocarbons (CFCs). While humans have little impact on the concentration of water vapor, there has been a significant increase in the concentrations of carbon dioxide and the other greenhouse gases. These increases have been the result of emissions from the burning of fossil fuels used by industry, homes, and cars.

A higher concentration of greenhouse gases results in more and more of the earth's outgoing energy being absorbed by the atmosphere. The higher amounts of absorbed radiation energy are converted into greater amounts of internal and molecular energy in the atmosphere. The atmosphere therefore radiates more energy to the ground below it. The ground and the air in contact with it therefore become warmer.

The term "greenhouse effect" is misleading. Greenhouses keep plants warm by trapping warm air inside and preventing it from mixing with the colder air outside. Greenhouse gases in the atmosphere, on the other hand, do not trap heat or radiation energy. They absorb this energy and convert it into other forms of energy that, through the processes of energy exchange between the atmosphere and ground, go into heating the earth.

Atmospheric scientists are busy trying to predict the effects that an increased greenhouse effect will have on the planet. Warming of the atmosphere could have a number of results. Vast agricultural areas may become too hot and dry to support plant growth. Polar ice may melt, leading to an increase in sea water levels, which could flood low-lying coastal areas. A warmer atmosphere may lend more energy to storms; hurricanes, for instance, may become stronger or more frequent in a warmer atmosphere. ***See also: Air; Atmosphere; Global Warming; Oceans; Radiation; Temperature; Water Vapor.***

HAIL

Chunks of ice that fall during a thunderstorm are called hail. Hailstones can vary from the size of large seeds to the size of grapefruits. Since hail is produced by thunderstorms, and since thunderstorms occur most often during the spring and summer, hail is a warm-season weather phenomenon. It is sometimes confused with sleet, a cold-season phenomenon that is produced under different circumstances. Hail causes hundreds of millions of dollars in damage annually in the United States, mostly to crops. Large hail can seriously damage cars and roofs of buildings, and even cause injury to anyone caught outside.

This car shows dents from large hailstones that fell in a storm in Fort Worth, Texas.

Hail forms in the violent updrafts and downdrafts inside a thunderstorm. These currents of air rise and fall alongside one another in a thunderstorm and contain an enormous quantity of tiny water droplets. The water droplets are supercooled, meaning that they remain in liquid form even at temperatures well below freezing.

These supercooled drops freeze when they collide with microscopic, airborne particles such as dust, or with ice crystals. These small ice particles are carried aloft by the thunderstorm's updraft. As they rise higher, more and more water freezes around the ice particle and a hailstone is born. The growing hailstone can remain suspended in the updraft, or it can rise and fall through the thunderstorm as it gets thrown from updraft to downdraft and back again. As the hailstone rises and falls through layers of warmer and colder air, it gains successive coatings of ice. The expanding coat of ice causes the hailstone to weigh more and more, until it becomes too heavy to remain suspended in air by the updraft. It then falls to earth. When larger hailstones are sawed in half, they often display rings of ice that form during these vertical ascents and descents, much like tree rings form during periods of growth and dormancy.

The stronger the thunderstorm's updraft, the heavier and the larger the hailstones that it can bear. Large hail, therefore, is usually indicative of an intense thunderstorm. Some of the larger hailstones can fall to the earth at speeds of 70 mph. The largest recorded hailstone in the world fell at Coffeyville, Kansas, in 1970. It weighed 1.67 pounds and was 5.5 inches in diameter.

Hail can occur anywhere, but in the United States it occurs most frequently in the Plains states. Moving thunderstorms can leave long swaths of hail in their wake. Sometimes hail-producing thunderstorms remain nearly stationary over a particular location. When this happens, a thunderstorm can dump so much hail that snowplows are required to clear roads, even in the middle of summer. *See also: Air; Sleet; Thunderstorms.*

Halo

A halo is a luminous ring circling the sun or moon. It is produced when light interacts with tiny ice crystals in the atmosphere. Halos sometimes appear twelve to twenty-four hours in advance of rain or snow, as thin icy clouds spread out ahead of an approaching storm.

Halos appear when sunlight or moonlight bends upon passing through ice crystals high in the atmosphere. Ice crystals frequently take the shape of a six-sided, or hexagonal, plate or column. These particular shapes tend to bend light at a 22° angle as it passes through.

A halo sometimes occurs when sunlight passes through tiny ice crystals in the sky.

Various kinds of halos, as well as other optical effects, can occur, depending on the variety and orientation of the ice crystals as light shines through them. The most common kind of halo occurs when light passes through large amounts of randomly oriented ice plates and columns. Since these crystals bend light at a 22° angle, a zone of bright, refracted light will appear in a circle around the sun at an angle of 22° from the viewer. A 46° halo occurs when light passes through ice crystals that are shaped like a hexagonal column, rather than a plate. When plate-like crystals are suspended in the air with their flat sides parallel to the ground, two luminous spots, called sun dogs, will appear on either side of the sun. Finally, a bright spot on a halo appearing directly above the sun or moon is called a tangent arc. This occurs when light passes through ice columns oriented with their long axis horizontal to the ground. *See also: Air; Atmosphere; Rain; Snow.*

HAZE

Haze consists of microscopic particles suspended in air. These particles reduce visibility and give a pale white or yellow, fuzzy appearance to the sky. Haze is a common occurrence around the world, though some locations experience more haze than others, owing to local geography and proximity to industrial areas. If haze is dense enough, it can be a health hazard as well as a hazard to small aircraft.

Haze forms as water vapor in the air interacts with tiny suspended particles called condensation nuclei. These particles can be pollutants from industry or cars, tiny salt particles from seawater, dust, pollen, or smoke. Many of these condensation nuclei attract water molecules, and are therefore called hygroscopic nuclei. Water vapor molecules in the atmosphere stick to these nuclei when they collide. These particles grow slightly larger as a thin film of water forms over them, and as they become visible they cast a milky white or yellow color across the sky. Objects in the distance become less visible as more and more haze forms in the atmosphere.

Haze can develop anywhere under certain conditions. Normally, wind keeps the air from accumulating much haze by blowing the haze and dispersing it through the atmosphere. However, when winds are light, haze can accumulate over an area. Slow-moving or stalled areas of high pressure, with their light wind flow, commonly result in a buildup of haze. Sometimes the local geography aids in this process, as when haze becomes trapped in a valley, preventing it from dispersing. Since haze is caused by water vapor interacting with microscopic nuclei, the more water vapor that exists in the air the more easily haze can form.

The Macmillan Encyclopedia of Weather

Haze can sometimes become so thick that it obscures visibility.

Thus, haze is most frequent in the summer when the air is more humid.

In the United States, haze is a frequent occurrence east of the Plains states. This is because air generally flows from west to east across the country. Relatively clear air will move onshore from the Pacific into the West Coast. As the air travels eastward over the country, it picks up more and more condensation nuclei. These haze-forming particles are particularly abundant across the eastern United States, downwind from industrial areas. Persons traveling between eastern and central or western parts of the United States can observe a dramatic change in the appearance of the sky due to the presence or absence of haze. In the East, clouds are often difficult to see in a haze-filled summer sky. In the Plains states, by contrast, skies are more often clear and blue, with clouds in sharp focus. In the West, haze can form in valley areas where air gets trapped and stagnates. Haze is less common in tropical regions, far removed from industry and pollutants. However, large forest fires in the tropics can cause vast amounts of haze that can drift for thousands of miles. ***See also: Air; Atmosphere; High Pressure; Visibility; Water Vapor.***

Heat Exhaustion and Heat Stroke

Hot weather, especially when accompanied by physical exertion, can inhibit the body's ability to cool itself, resulting in heat exhaustion or heat stroke. These health conditions occur when heat builds up in the body faster than it can be diminished by the body's natural cooling mechanisms. The heating and breakdown of the body's circulation causes organs to become damaged and malfunction. Heat stroke is the more dangerous of the two health conditions and can result in death if not treated promptly. Both heat exhaustion and heat stroke can affect anyone, though the elderly and ill are most at risk.

Heat exhaustion is the first illness that occurs when too much heat builds up in the body. Symptoms include elevated body temperature, faintness or nausea, rapid heartbeat, and low blood pressure. The person can have a pale appearance with cold, clammy skin. A person showing these signs should be moved to a cool area immediately. Clothing should be loosened or removed, and the person should lie down with their legs elevated. Cool, not cold, water should be consumed.

Heat stroke is the second stage in a heat-related illness and is much more serious. Symptoms include a body temperature of 105°F or higher, hot dry skin, a rapid heartbeat and irregular blood pressure, rapid and shallow breathing, and mental confusion. A person suffering from heat stroke should be given professional medical attention immediately. ***See also: Heat Index; Heat Wave; Temperature.***

Heat Index

The heat index, also known as the apparent temperature, factors the combination of temperature and humidity to measure how warm the air actually feels. The more humid the air, the warmer it feels, and the higher the heat index. The heat index is used in summer when air temperature and moisture levels become high enough to pose a hazard to health.

The amount of moisture in the air determines how warm the air feels by affecting evaporation rates. Evaporation occurs when water transforms from a liquid state to a gaseous state. The body uses evaporation to regulate temperature and keep cool in warm weather. When humans perspire in warm weather or after exercise, tiny water droplets form on the surface of the skin. As these drops evaporate, they remove heat energy from the air, causing the air near the skin to become cooler. This cooler layer of air near the skin helps regulate body temperature.

When the air is more humid, it contains more water vapor. This water vapor condenses, or changes from gaseous to liquid state, onto the skin. As it does, it releases heat energy into the surrounding air, thereby competing with the evaporation process. The more water vapor in the air, the harder it is for perspiration to evaporate and cool the air around the body.

This table shows the resulting heat index value based on temperature (far left-hand column) and humidity (top row).

One of the body's cooling mechanisms is perspiration. When the body sweats, tiny water droplets collect on the surface of the skin and evaporate into the air. Since the process of evaporation requires energy, when sweat evaporates from the skin it removes heat energy from the surrounding air and enables the body to cool. The perspiration process is not as efficient in the elderly. This is why the elderly are at a greater danger of heat exhaustion by exposure to hot weather, even when they are not exerting themselves physically.

Heat Index Chart - National Weather Service

Temperature °F \ Relative Humidity %	0	5	10	15	20	25	30	35	40	45	50	55	60	65	70	75	80	85	90	95	100
70	64	64	65	65	66	66	67	67	68	68	69	69	70	70	70	70	71	71	71	71	72
75	69	69	70	71	72	72	73	73	74	74	75	75	76	76	77	77	78	78	79	79	80
80	73	74	75	76	77	77	78	78	79	79	80	81	81	82	83	85	85	86	87	88	89
85	78	79	80	81	82	83	84	85	86	87	88	89	90	91	93	95	97	99	102	105	108
90	83	84	85	86	87	88	90	91	93	95	96	98	100	102	106	109	113	117	122		
95	87	88	90	91	93	94	96	98	101	104	107	110	114	119	124	130	136				
100	91	93	95	97	99	101	104	107	110	115	120	126	132	138	144						
105	95	97	100	102	105	109	113	118	123	129	135	142	149								
110	99	102	105	106	112	117	123	130	137	143	150										
115	103	107	111	115	120	127	135	143	151												
120	107	111	116	123	130	137	148														
125	111	116	123	131	141																
130	117	122	131																		
135	120	128																			
140	125																				

The National Weather Service began using the heat index in 1984 as a way to communicate the dangers of temperature and high humidity levels to the public. In the summer, when temperatures rise into the 90s, a high humidity level can make the temperature feel much hotter. Heat indices above 100° may result in health hazards such as heat exhaustion or heat stroke. ***See also:* Heat Exhaustion and Heat Stroke; Heat Transfer; National Weather Service; Temperature; Water Vapor.**

Heat Transfer

When a substance is heated, the average motion of its molecules increases. This increase in energy is measured by the object's temperature. There are three ways that energy can be transferred between objects with different temperatures: conduction, convection, and radiation. The amount of heat in the air, as measured by the temperature, is a result of all three heat transfer processes.

The air can be heated by contact with warm ground through the process of conduction. This occurs when faster moving molecules in a warmer object collide with slower moving molecules in another object. The faster moving molecules will lose some of their speed, or kinetic energy, to the slower molecules, thereby causing the warm object to cool. The slower molecules, on the other hand, will gain kinetic energy and speed up, thereby causing the cooler object to become warmer.

Heat is also transferred through the air via convection. This occurs when currents of warmed air rise into the sky and spread the heat vertically through the atmosphere. Convection is the transfer of heat by the movement of large amounts of molecules in a gas or fluid, such as air or water. A volume of warm air near the earth's surface expands and becomes less dense, causing it to rise. As the air rises and expands, it cools. If the temperature drops low enough, water vapor molecules in the air will condense into liquid water droplets and a cloud will form. The process of convection results in the small, puffy cumulus clouds that frequently dot the summer sky.

The third heat transfer process is radiation. The sun and the atmosphere radiate energy to the ground and the oceans, causing them to become warmer. All objects with molecular motion emit electromagnetic waves, called radiation. These waves, which can also be thought of as individual packets of energy called photons, travel at the speed of light. When an object absorbs incoming radiation, it gains energy. Some of this additional energy causes the molecules in the object to move faster, resulting in an increase in the object's temperature.

Radiation affects the temperature of the air in two ways. The first way occurs when the ground radiates energy into the atmosphere. Some gas molecules in the atmosphere absorb this energy. These energized molecules move around faster, on average, thereby increasing the temperature of the air. The second process is less direct. As the radiated energy from the sun warms the ground and oceans, these surfaces increase the temperature of the air through conduction. *See also: Atmosphere; Cloud Formation; Oceans; Radiation; Temperature; Water Vapor*.

If heat wave conditions persist and are accompanied by drought, vegetation can be severely damaged.

Heat Wave

An extreme example of a heat wave made worse by high humidity levels occurred in the Midwest and parts of the East Coast of the United States in July 1995. Hot air gathered over parts of the Midwest that had recently experienced heavy rain. The hot air became oppressively humid, with moisture readings that were more typical of tropical oceans than of the central United States. In Chicago, temperatures in the upper 90s and low 100s combined with the excessive humidity made the air feel like the temperature was over 120°F. As these oppressive conditions continued for over a week, over 600 people died. Most were elderly or sick people confined to homes or apartments without air conditioning. The hot, humid weather then spread into parts of the Northeast, causing even more fatalities.

A heat wave is a prolonged period of temperatures well above average for the season. Most heat waves last anywhere from a few days to a week or more, usually in the summer when warm air is more widespread. Heat waves can pose serious risks to health and are sometimes accompanied by drought, which can cause major damage to crops.

One way heat waves occur is when warm air builds up over a part of the country during a stagnant weather pattern. Stagnant weather conditions often occur in conjunction with areas of high pressure, which are accompanied by light winds that slowly circulate in a clockwise direction. If the high pressure moves slowly or stalls, heat can build over a region for days and temperatures can rise well above average. As heat builds up underneath a high-pressure system, southerly winds circle around the western edge of the high-pressure area and can carry the heat northward.

An important factor in many heat wave events is the role of topography. When winds blow warm air from higher elevations to lower elevations, it can cause the air to become even warmer. As air descends a slope, more and more of the atmosphere weighs down from above. This exerts higher and higher pressure on the air and compresses its volume. When air compresses, the molecules in the air move faster and collide more often. This increasing molecular energy causes the temperature to rise. The greater the elevation difference between where the air begins and where it ends, the more warming can occur.

In the northeastern United States, the hottest summer days occur when winds blow from the west or northwest, rather than the south. This is because the west to northwest winds descend the Appalachian Mountains, causing air to warm considerably by the time it reaches the large coastal cities from Boston to Washington, D.C. Southerly winds do not descend any elevation, and in many parts of the northeast a southerly wind blows in from the relatively cooler Atlantic Ocean. In the western United States, on the other hand, the hottest days occur when winds blow from the east. Easterly winds descend thousands of feet from the interior mountain chains to coastal areas, resulting in very hot and dry conditions. When strong winds accompany this hot, dry easterly flow of air, conditions become favorable for wildfires.

Another factor in heat waves is soil moisture. The drier the soil, the more the sun's energy will go into heating the soil rather than evaporating soil moisture. As a result, drier soil becomes much hotter and transfers more heat energy to

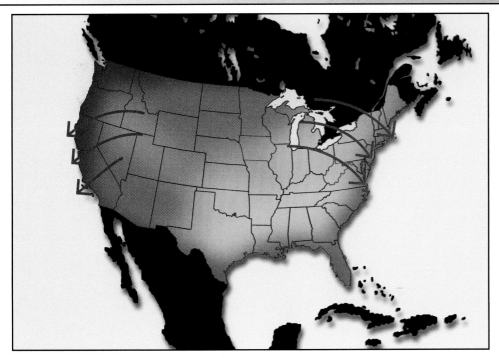

On the west coast of the United States, heat waves are often accompanied by easterly winds. On the east coast, westerly or northwesterly winds often accompany heat waves.

the air in contact with it. Moist soil, however, can cause heat waves to become dangerous in a different way. Hot air that moves over an area where soil is particularly wet from recent rains will cool slightly but will become very humid. Humidity can make a heat wave even more hazardous to humans by hindering the body's ability to cool itself through perspiration. *See also: Atmosphere; Drought; Heat Transfer; High Pressure; Oceans; Temperature.*

HIGH PRESSURE

A high-pressure area, or "high," is a large atmospheric circulation (known as an anticyclone by meteorologists), the center of which has a higher atmospheric pressure than all surrounding areas. Winds spiral clockwise and outward around high-pressure systems. Highs are mostly accompanied by fair and precipitation-free weather.

THE LIFE CYCLE OF A HIGH-PRESSURE SYSTEM

High-pressure areas generally form underneath areas of converging wind at high elevations in the atmosphere. A converging wind brings air together, increasing the mass of air over an area. The more air that converging winds bring into an area high in the atmosphere, the more air there is to weigh down on the surface of the earth and the higher the air pressure becomes. Since air cannot pile up over a region indefinitely, some of the air begins to sink toward the earth. Once it reaches the ground, it spreads apart horizontally, or diverges. A diverging wind spreads air apart, decreasing the mass of air. As long as more air converges aloft than diverges at the ground, air pressure will rise and the high-pressure area will strengthen, or build.

Areas of upper-level convergence are usually associated with large-scale weather systems. As the weather systems move along, so do the areas of converging air aloft and the high-pressure systems underneath them. If the magnitude of upper-level convergence weakens, so will the area of high pressure.

Effects of High-Pressure Systems

The sinking air in high-pressure areas is responsible for the fair weather that these systems frequently produce. Clouds form when rising air causes water vapor to change from gaseous molecules to liquid droplets. Sinking air, however, causes water droplets to evaporate into gaseous water molecules, thus inhibiting cloud formation.

The weather around a high-pressure area differs depending on the location. Since the wind blows in a clockwise, or anticyclonic, direction around a high-pressure area, locations on the eastern side of a high-pressure system will experience a general northerly flow of air. This can often result in cool, dry weather. On the western side of a high-pressure area, winds blow from the south. Thus, areas to the west of the center of a high often experience warm, and sometimes humid, weather. *See also: Atmosphere; Cloud Formation; Convergence; Precipitation; Water Vapor; Wind.*

Hurricanes

A hurricane is a giant atmospheric circulation that forms over warm tropical ocean water. An organized hurricane has a calm, clear eye at the center of the system where winds are light and air pressure is lowest. The area of violent storminess around the eye is called the eye wall, and that is where the strongest winds blow. To be classified as a hurricane, a storm must have winds of 74 mph (65 knots) or higher, and the strongest hurricanes have winds of 150 mph or more. Hurricanes can be as wide as 400 miles in diameter and can last for days or weeks.

The Structure of a Hurricane

A hurricane, while consisting of violent winds and rain, maintains a very organized circulation of air. Near the surface of the earth underneath a hurricane, air

From space, a hurricane looks like a giant swirling cloud pattern.

spirals in toward the center of the system in a counterclockwise direction. This air picks up heat and moisture from the warm ocean water. As the warm air rises and cools, the tropical moisture condenses and changes from gaseous to liquid state. This forms billions of water droplets, which create the clouds and thunderstorms inside a hurricane. This process of condensation also gives off heat energy, which fuels the hurricane.

As air spirals closer to the center of the hurricane, it blows faster and faster and rises more and more violently. Rain becomes torrential. The strongest winds and heaviest rain occur in the eye wall, near the center of the storm. Eyes of hurricanes range anywhere from five to 120 miles in diameter, but they average about thirty miles in diameter. While air spirals inward and rises inside a hurricane, air actually sinks inside the hurricane's eye. Sinking air causes moisture in the air to evaporate and change from liquid water to invisible gaseous water vapor. Therefore, in well-developed hurricanes, the eye of the storm will often be sunny and free of clouds.

Outside of the eye, rising air inside a hurricane eventually spreads out horizontally as it reaches high elevations in the atmosphere. This horizontally spreading air circulates away from the center of the hurricane in a clockwise direction. This outward circulation at high elevations over the hurricane is said to diverge because it removes air from over the system. As air is removed, more and more air must rise from below to take its place. Therefore, the outward spreading air at high elevations over the hurricane acts as a kind of ventilation system.

The Macmillan Encyclopedia of Weather

There are several important factors that must be present for a cluster of thunderstorms to eventually develop into a hurricane.

- The ocean water temperature must be 80°F or higher. This warm water serves as a source of heat and moisture, which fuels the development of hurricanes.
- The air temperature should decrease relatively rapidly with increasing height. Such an atmosphere is said to be unstable because it favors rising air. Rising air contributes to the formation of clouds and thunderstorms.
- The atmosphere should be relatively humid. Clouds and thunderstorms that comprise a hurricane are made from atmospheric moisture. Dry air limits the amount of clouds and storms that can form.
- The cluster of thunderstorms must be at least 300 miles away from the equator. This is because hurricanes get their spin from a force caused by the rotation of the earth. This force is equal to zero at the equator and increases with increasing latitude.

A cross section of a hurricane shows how air flows into the center of the storm at low levels, rises up through the storm, then flows outward at high levels.

While this describes the circulation pattern found in an organized hurricane, some hurricanes are less organized and can have a somewhat different structure. A hurricane can lack organization if winds in upper levels of the atmosphere increase and blow over the top of the system. This prevents air from diverging and ventilating the top of the hurricane. If these disruptive winds are strong enough, they can push all the clouds and storms to one side of the hurricane, causing the eye of the storm to disappear and exposing the center of the storm. Hurricanes can completely dissipate if strong upper-level winds blow over the top of the system for several days.

DEVELOPMENT OF HURRICANES

Hurricanes are steered by the flow of air through the atmosphere. They die when they move over colder water or land. In the Northern Hemisphere, hurricane activity is at its peak when ocean water is at its warmest in the late summer and early fall. Even so, hurricanes have been recorded as early as June and as late as December.

The clusters of thunderstorms from which hurricanes arise form over tropical oceans from a couple of sources. Sometimes a weakening cold front will move in from the north and stall over warm ocean water. Showers and thunderstorms along the front occasionally will merge together, strengthen, and develop a circulation that eventually will give rise to a hurricane. More frequently, however, groups of thunderstorms occur in association with atmospheric disturbances called tropical easterly waves. Easterly waves move from east to west through the tropics with the normal easterly flow of air.

These easterly waves occur in a zone of showers and thunderstorms that stretches like a belt around the planet near the equator. This zone is called the

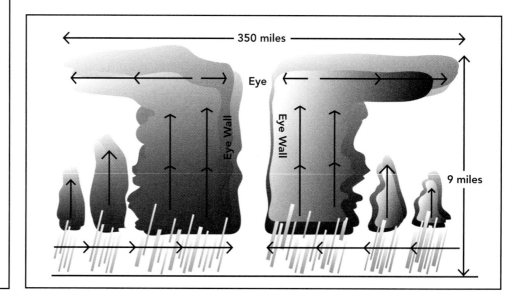

intertropical convergence zone, or ITCZ, by meteorologists. The ITCZ is an area where two different flows of tropical air, one coming from the northeast and one from the southeast, come together and converge. As the air collides, it piles up and rises into the atmosphere. As the rising air cools, abundant tropical moisture in the air transforms into liquid water droplets. This forms clouds, showers, and thunderstorms. Easterly waves form in the ITCZ, and some of these easterly waves go on to become hurricanes.

The exact process whereby a group of thunderstorms develops the circulating winds that eventually spin up into a hurricane is still not fully understood. However, meteorologists have observed that large clusters of thunderstorms naturally tend to produce counterclockwise circulations about three miles above the earth's surface. As the thunderstorms go through cycles of development and weakening over a period of several days, this circulation can strengthen. When the counterclockwise air circulation develops in low levels of the atmosphere, near the ocean, and the thunderstorms increase and become more organized, a tropical depression forms. As the circulation increases, winds strengthen. The system is classified as a tropical storm when winds reach 39 mph (35 knots) and the system is assigned a name from a predetermined list. When the winds reach 74 mph (65 knots), the system is classified as a hurricane.

Hurricanes move very much like a floating object in a river. In the case of a hurricane, the "river" is the flow of air from the surface of the earth up to a height of around 15,000 feet. In the tropics, this river of air moves out of the east. Therefore, hurricanes generally move from east to west in the tropics with a slight northward movement. As the hurricane gets farther and farther north, the direction of the atmospheric steering currents changes to westerly. As a result, hurricanes generally tend to turn toward the north, then to the northeast as they get caught in this westerly flow of air. Circulation patterns around large-scale weather systems, however, can steer hurricanes in many directions. They can stall and even perform a loop.

- *Winds should not change much in direction or speed with increasing height. The more wind changes with height the greater the shear in the wind. Wind shear disrupts developing hurricanes by blowing apart a hurricane's thunderstorms.*
- *Winds at high levels of the atmosphere should spread apart, or diverge. Diverging winds remove air from over a developing hurricane. This allows more air to rise up and take its place. The formation of the clouds and thunderstorms of a hurricane is due to rising air.*

Since so many different weather conditions must be favorable for a cluster of thunderstorms to develop into a hurricane, most thunderstorm activity and easterly waves form and die in the tropics without ever strengthening.

EFFECTS OF HURRICANES

Hurricanes threaten life and property in several ways. When over the ocean, high winds and immense seas threaten any ships that stray into or near a hurricane. When a hurricane moves over land, winds can cause significant damage, and torrential rains can result in flooding. Hurricanes also occasionally spin off tornadoes as they move inland. Historically, however, the deadliest effect of a hurricane is

Hurricane Wind Speed (mph)	Storm Surge Height (feet)
74–95	4–5
96–110	6–8
111–130	9–12
131–155	13–18
156 +	19 +

A satellite photo clearly shows the calm eye of a hurricane.

its storm surge. Storm surges occur as hurricane winds pile up ocean water along the coast, resulting in major flooding of low-lying coastal areas.

A storm surge forms as a result of the action of strong winds and ocean water. Winds blow in a counterclockwise circulation around a hurricane. As a hurricane moves toward a coastline, therefore, winds on the right-hand side of the storm (with respect to the storm's movement) will blow toward the coast. This strong, onshore-directed wind flow will cause ocean water to pile up along the coastline. Near the center of the storm, where the winds are strongest, ocean water will rise many feet above the normal tide level. Hurricane winds push this ocean water inland as the storm moves onshore, flooding low-elevation coastal locations.

The magnitude and destructive power of a storm surge depends on the strength of the storm, the coastline, and the tide level. The stronger the storm, the faster the winds will blow and the more water will pile up along the coast. Coastlines that curve inward, toward land, will tend to funnel storm surges into a narrower area and pile up water even higher. Areas of coastline where the ocean bottom is relatively shallow will also cause water to pile up higher. And if a hurricane moves across the coast at the time of normal high tide, the effects of a storm surge will be much worse than if the hurricane moves inland during low tide. Storm surges range from two to four feet in weaker hurricanes to as high as thirty feet in rare instances during particularly intense hurricanes.

Winds in a hurricane not only pile up water along the coast but can cause structural damage on their own. Because winds must blow at least 74 mph for a sustained period in the weakest hurricane, winds this strong can easily blow over trees and cause other minor damage. In the strongest hurricanes, sustained winds can blow at 150 mph or more with gusts near 200 mph. Winds such as these are equivalent to a strong tornado and can destroy buildings.

Hurricanes drop enormous amounts of rain; five to ten inches of rain may easily fall on a given location. If the hurricane moves more slowly, even more rain can fall—sometimes twenty inches or more. Flooding will often continue well away from the coast as a hurricane moves inland and dissipates.

When hurricanes move inland, they often spin off tornadoes. Tornadoes are most likely to occur on the north and east sides of a hurricane, in thunderstorms that form well away from the center of the storm. Most hurricane-spawned tornadoes are weak, but occasionally a hurricane will produce a strong and damaging tornado.

The Saffir-Simpson scale measures hurricane strength and damage potential. Hurricanes are ranked on a scale from one to five, with five being the strongest. Only two category five hurricanes have hit the United States in the last 100 years.

Hurricanes: Historical Events

On average, around six hurricanes form in the Atlantic Ocean, including the Caribbean Sea and Gulf of Mexico, each year. An average of two of them will become "intense," meaning they will have sustained winds of 110 mph or more. When these intense storms encounter land, they can cause immense destruction and loss of life. Some of the more destructive hurricanes of the twentieth

Destructive Hurricanes of the Twentieth Century

Location	Year	Highest Winds (mph)	Deaths	Damage	Comments
Galveston, TX	1900	115+	8,000+	$800 million	Deadliest U.S. Hurricane
Florida Keys	1935	155+	400+		Lowest barometric pressure recorded in U.S.
New England	1938	80, gusts 100+	600+	$306 million	Most destructive hurricane ever in Northeast U.S.
Mississippi (Camille)	1969	180, gusts 20+	250	$1.4 billion	Thirty foor storm surge along Mississippi coast; massive flooding inland
South Carolina (Hugo)	1989	130	app. 50	$7 billion	Second costliest hurricane in U.S. history
South Florida Louisiana (Andrew)	1992	140	40+	$26.5 billion	Costliest hurricane in U.S. history
Hawaii (Iniki)	1992	140	6	$2 billion	Most destructive Hawaiian hurricane ever.

This table lists some of the most destructive hurricanes of the twentieth century.

century are listed below; note that prior to 1953, wind speeds were not methodically recorded and hurricanes were not named.

FORECASTING HURRICANES

Meteorologists forecast the development, movement, and changes in strength associated with tropical depressions, tropical storms, and hurricanes once they form. Other researchers also try to forecast well ahead of time how many hurricanes will develop during the hurricane season, which runs from June through November.

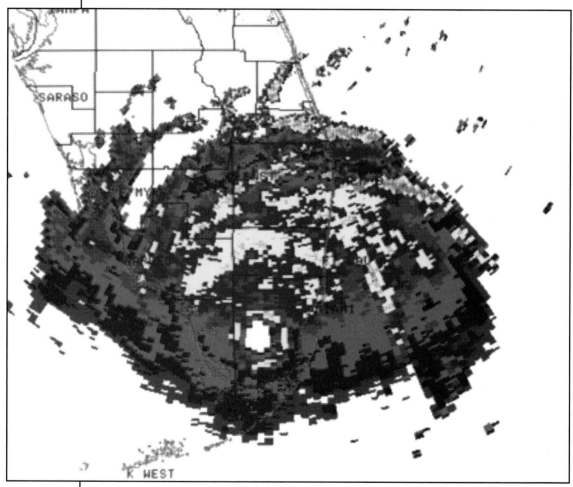

Meteorologists use radar to track hurricanes as they move onshore.

Forecasting hurricanes is extremely difficult. Meteorologists use computers, satellites, and other tools to help them. In the United States, the National Hurricane Center in Miami, Florida, is charged with tracking and forecasting tropical depressions, storms, and hurricanes throughout the year across the Atlantic and eastern Pacific Oceans.

Meteorologists use a variety of forecasting methods when predicting the development and movement of tropical systems. One method is called climatology and is based on long-term averages of weather. When meteorologists look at past weather events to see what has happened in situations similar to the one they are forecasting for, they are using climatology. Another method is called persistence. A persistence forecast is one that presumes the movement of a hurricane or tropical weather system will persist in the same direction and speed that it has been moving over the past day or two. Meteorologists also use supercomputers, on which they run extremely complex software that simulates the atmosphere and comes up with mathematically-formulated predictions of hurricane movement and intensity. Mixed into all of these forecasting methods is human intuition. All weather forecasters use their own experiences and understanding of atmospheric processes coming up with a prediction of where a storm is headed.

A number of tools are at the meteorologists' disposal in helping them arrive at a forecast. Satellites take photographs that track hurricanes from their beginnings as disorganized clusters of thunderstorms to their peak intensity. Satellite imagery is particularly useful in monitoring remote parts of the oceans where no islands and few ships exist to take direct weather observations. Once a hurricane approaches land, radar stations along the coast can directly monitor the movement and internal structure of a storm.

An extremely important tool that hurricane forecasters use is aircraft. In the Atlantic, when a hurricane comes within flying distance of islands or other land areas, instrument-laden aircraft are dispatched to fly into the storm and take direct weather measurements. The people that undertake these dangerous missions are called hurricane hunters. They provide vital information about exactly how strong a hurricane is and exactly where it is located.

Outside of hurricane season, well away from the tropics, a group of researchers at Colorado State University each year make long-range predictions about the number of hurricanes that will form in the Atlantic during the hurricane season. They forecast whether hurricane activity will be above or below the long-term average and how many intense hurricanes with winds over 110 mph will form. Their predictions are based on several factors. One is rainfall in western Africa. Some of the atmospheric disturbances that go on to become hurricanes have their origins over western Africa. These disturbances are more likely to form when rainfall in this area is above normal. Another factor these researchers consider is whether El Niño is present or not. El Niño is a warming of sea surface temperatures in the eastern tropical Pacific Ocean. One of the effects of El Niño is increased wind speeds in upper levels of the atmosphere that blow over the tropical Atlantic. Strong upper-level winds shear developing hurricanes apart. Research has shown that when a strong El Niño is present, hurricane activity will be much reduced in the Atlantic Ocean.

When a hurricane approaches the coast, the National Hurricane Center will issue watches or warnings. A tropical storm watch means that tropical storm conditions, with winds of 39 mph or more, are possible in the watch area within thirty-six hours. A tropical storm warning means tropical storm conditions are expected in the warning area within twenty-four hours. Likewise, a hurricane watch means hurricane conditions are possible in the watch area within thirty-six hours. A hurricane warning means those conditions are expected within twenty-four hours. All preparations should be rushed to completion, and coastal areas within the hurricane warning should be evacuated. *See also: Atmosphere; Cloud Formation; Easterly Waves; El Niño; Heat Transfer; Intertropical Convergence Zone; Oceans; Radar; Rain; Storm Surge; Thunderstorms; Tornadoes; Satellite; Water Vapor; Wind.*

HYDROLOGICAL CYCLE

The hydrological cycle describes how liquid and gaseous water circulate among the atmosphere, land, and the ocean. It involves the processes whereby water changes state from gaseous water vapor to liquid water or solid ice, and back again.

The atmosphere, with its rainfall and snowfall, plays a necessary link in the hydrological cycle. There are two basic steps in the hydrological cycle: the movement of water from the surface of the earth into the atmosphere, and the movement of water from the atmosphere to the surface of the earth. Water also circulates from land areas to the ocean.

In the atmosphere, water vapor rises on vertical air currents. This rising motion, called convection, spreads moisture high into the sky. As air rises higher it expands and cools. The water vapor molecules in the cooling air move slower and slower, and instead of remaining separate they begin to adhere when they collide with each other or with tiny airborne particles. Water vapor molecules that stick together like this become tiny water droplets. This process, whereby water vapor changes into liquid water droplets, is called condensation and is the opposite of evaporation. Condensation is the process that forms clouds in the atmosphere. If enough condensation occurs, a cloud can grow large enough to produce rain or snow—called precipitation by meteorologists.

Precipitation transfers water back to the surface of the earth from the atmosphere. Some of the precipitation that falls evaporates before reaching the ground, thereby becoming water vapor again. Precipitation that reaches the ground runs off into streams, rivers, and lakes. Runoff is also absorbed by soil

Streams and rivers carry water back to the ocean as part of the hydrological cycle.

For water to move from the surface of the earth upward into the atmosphere, evaporation must occur. Evaporation is the process whereby liquid water changes to gaseous water vapor. It occurs as water vapor molecules fly apart and escape from a liquid water surface. This process is continually occurring at the surface of lakes, puddles, rivers, and oceans. Vast quantities of water evaporate from the surface of the earth into the atmosphere.

and used by plants, where it is released as water vapor back into the atmosphere through a process called transpiration. Runoff from rain or snow drains into streams, lakes, and rivers, which eventually flow back into the ocean. The hydrological cycle is therefore a continuous exchange of water between the atmosphere and the ground, with no beginning or end. *See also: Air; Atmosphere; Cloud Formation; Oceans; Precipitation; Rain; Snow; Water Vapor.*

HYDROSTATIC EQUILIBRIUM

Hydrostatic equilibrium describes the balance between forces that act in vertical directions in the atmosphere. It states that the force exerted by the vertical difference in air pressure is counterbalanced by the force of gravity. Hydrostatic equilibrium explains why the earth's atmosphere does not fly off into space and why atmospheric air currents are much stronger in a horizontal direction than they are in a vertical direction.

One of the two forces in hydrostatic equilibrium is the force that is generated by differences in air pressure. This force is called the pressure gradient force.

The pressure gradient force tends to move air from high pressure to low pressure. Since air pressure in the atmosphere is highest near the earth and lowest at high altitudes above the earth, an upward directed force is exerted on the air. This force, in the absence of any other, would cause the atmosphere to rush off into space. Gravity, which exerts a force that pulls down on air molecules, counterbalances this upward directed pressure gradient force. Horizontal pressure differences across the earth's surface, on the other hand, are not counterbalanced by any force. Therefore, air moves horizontally around the planet much more quickly than it moves vertically in the atmosphere.

Hydrostatic equilibrium is a generalization. Local variations in temperature and air pressure cause the hydrostatic equilibrium to become slightly out of balance, so air rises or falls. Vertical air currents are required in the atmosphere to distribute heat and moisture and to create the clouds that produce rainfall. *See also: Atmosphere; Cloud Formation; High Pressure; Low Pressure; Temperature*.

HYGROMETER

A hygrometer is a device used to measure the humidity of the air. Two kinds of hygrometers are a hair hygrometer and an electrical hygrometer.

A hair hygrometer is based on the behavior of hair as it comes into contact with moisture in the air. Hair absorbs airborne moisture. The more humid the air, the more moisture hair absorbs. As hair absorbs moisture it expands by as much as 2 to 3 percent in length. In a hair hygrometer, several strands of hair are attached to a mechanical pointing device. As the hair expands and contracts because of the changing humidity levels, the pointer moves up and down on a chart, which indicates the air's relative humidity.

Hair hygrometers require frequent adjustment and do not respond well in cold weather. Scientists therefore use electrical hygrometers to measure humidity levels. In an electrical hygrometer, a flat metallic plate is coated with a layer of carbon. An electric current that runs across the plate varies in strength, depending on the amount of atmospheric moisture that collects on the plate. The flow of the current, therefore, is electronically transformed into a measurement of humidity.

A relative humidity reading of 100 percent indicates that water vapor molecules are changing into liquid water more rapidly than liquid water is changing into gaseous water vapor. At ground level, 100 percent humidity is always accompanied by fog. A relative humidity reading near zero, on the other hand, indicates that water vapor molecules are only very infrequently changing into liquid water molecules, and then only for a brief moment before flying apart as water vapor again. This occurs when a low amount of water vapor is in the air. The relative humidity on a hygrometer can never read zero except in a controlled laboratory environment because there is always at least a small amount of water vapor molecules in the air. *See also: Air; Fog; Water Vapor.*

Hypothermia

Hypothermia is a dangerous and life-threatening health condition that occurs when the body temperature drops below 95°F. Hypothermia occurs after prolonged exposure to cold, when heat is drained from the body faster than the body can replace it. It can lead to death, most often when people are exposed to frigid air during wintertime cold waves.

Symptoms of hypothermia include a general physical and mental deterioration. The first signs of hypothermia include drowsiness and irritability, as well as mental confusion and slurred speech. The person may become pale, with slow and weak breathing and a low heart rate. If not treated, hypothermia will eventually lead to coma.

Hypothermia requires immediate medical attention. The body should be warmed slowly rather than quickly. This warming process is delicate and is best performed in a medical facility. To prevent hypothermia, people should limit their exposure to extreme cold. Clothing should be appropriate to the season and worn in layers. Head and neck protection are particularly effective in preventing heat from escaping the body. For the hands, mittens retain heat better than gloves. Alcohol should be avoided, as it increases the body's risk of hypothermia. *See also: Cold Wave; Temperature.*

Ice Age

An ice age is a period in the earth's history characterized by low temperatures and expansion of glaciers and polar ice caps. Ice ages typically consist of a cool period lasting for tens to hundreds of millions of years. Within this cool period, an even colder period will occur, lasting for tens of thousands of years. It is during this colder period that great ice sheets spread far southward across the continents. There have

During an ice age, glaciers expand to cover large sections of continents.

been several ice ages throughout the earth's history, but the term is most often applied to the most recent ice age, which began two to three million years ago and ended around ten thousand years ago. Ice ages are natural processes caused by climatic changes occurring on a planetary scale.

One reason ice ages come and go over millions of years is that the earth's orbit around the sun slowly changes. Changes in the earth's orbit result in significant variations in the amount of sunlight that reaches the earth. Since the sun's energy drives the earth's temperature, changes in the amount of solar energy can cause major swings in global climate.

There are three main ways in which the earth's orbit changes over time. The first involves the path that the earth takes as it travels around the sun. The earth's orbit changes from being nearly circular to more elliptical over a period of about 100,000 years. The second way the earth's orbit changes relates to the tilt of the earth on its axis. Rather than sitting exactly vertical with respect to the sun, the earth is tilted at an angle of about 23.5°. Over a period of around 41,000 years, this tilt changes by a couple of degrees. The earth's tilt not only changes its angle but also the direction in which it points. This directional change happens over a time span of about 11,000 years. Since these variations in orbit occur on different time scales, their effects can compete or combine to change the amount of the sun's energy that reaches the earth.

Another reason ice ages occur is the process of continental drift. The theory of plate tectonics describes how the earth's crust is made up of moving sections, much like a jigsaw puzzle. These plates constantly move against each other, spread apart from one another, or dive underneath each other. While their yearly rate of movement is slow, only a few centimeters, over millions of years this results in large changes in position. Areas of the world that today are located near the poles and covered in snow and ice were much closer to the equator millions of years ago. The southward expansion of polar ice sheets during an ice age requires large land areas at high latitudes over which the ice can travel. When moving plates distribute land areas into polar regions, ice ages are more likely to occur.

Most scientists agree we are still emerging from the most recent ice age, which peaked about 18,000 years ago. This poses questions about the atmospheric warming currently being observed around the planet. How much of the earth's current global warming is a natural result of the planet proceeding out of

Earth's orbit around the Sun changes over millions of years.

the recent ice age, and how much global warming is the result of human-caused changes to our atmosphere? Scientists are studying these questions in an effort to better understand how the global climate system works. *See also: Climate; Global Warming; Snow; Temperature.*

INTERTROPICAL CONVERGENCE ZONE

The intertropical convergence zone, or ITCZ, is a belt of scattered cloudiness, showers, and thunderstorms that stretches around the planet near the equator. It exists as a result of the way in which air circulates around the globe, and it is known by meteorologists as a breeding ground for atmospheric disturbances that develop into hurricanes. The ITCZ also contributes to the humid, rainy climate that supports tropical rain forests.

The intertropical convergence zone causes drenching rain showers across tropical regions.

The intertropical convergence zone is so named because it is the zone where two flows of air, one from the northeast and one from the southeast, converge. These flows of air come together in the tropics because the tropics are, on average, the warmest part of the earth. As the sun warms the surface of the earth across the tropics, the air in contact with the earth also warms. The warmer air becomes less dense and rises into the atmosphere. Wind blows in horizontally from the northeast and southeast to replenish this rising air in the tropics.

What Happens in the ITCZ

Rising air along the ITCZ contributes to the formation of clouds, showers, and thunderstorms. When air rises, it moves into lower and lower atmospheric pressure. The rising air therefore expands, the air molecules slow down, and the air temperature cools. Gaseous water vapor molecules in the air begin to turn into water droplets as they adhere to one another or to microscopic particles. This process, called condensation, is what forms clouds that produce rainfall.

The converging and rising air along the ITCZ is one part of a vast atmospheric circulation that extends across thousands of miles. As rising air along the ITCZ reaches high elevations in the atmosphere, it spreads out horizontally toward the poles. This poleward spreading air eventually sinks back to the earth at about 30° north and south latitude, and flows back toward the ITCZ.

ITCZ AND THE FORMATION OF HURRICANES

During the summer and early autumn in the Northern Hemisphere, when the ocean is warmest, some of the thunderstorms in the ITCZ organize and expand, developing their own circulation. If conditions are favorable, these thunderstorm clusters may eventually develop into hurricanes. The hurricanes then spin toward the west, eventually hitting land or curving back to the north and east and dissipating thousands of miles from where they emerged. From June to November, therefore, meteorologists constantly monitor the ITCZ for signs of developing hurricanes. *See also: Atmosphere; Climate; Cloud Formation; Convergence; Oceans; Thunderstorms; Rain; Water Vapor; Wind.*

ISOPLETH

Isopleths are lines that connect equal data points on a weather map. Most people have seen isopleths that connect locations of equal elevation on maps showing topography. Meteorologists use isopleths on maps every day to analyze patterns of temperature, moisture, wind, and pressure. Isopleths help the data stand out; a meteorologist can see areas of minimum and maximum values, and can easily identify areas where the weather changes rapidly from one location to another.

Isopleths are similar to connect-the-dot drawings. They can be drawn by hand or by a computer. When drawing isopleths on a map of temperature, for example, the 80° isopleth will be drawn connecting cities with a temperature of 80°. When drawing isopleths on weather maps, however, a certain amount of guessing must be done. This is because meteorologists do not have weather observations at every point across the country; rather, readings of temperature, pressure, moisture, and wind are taken at widely separated locations. When these weather observations are plotted on a map, there are large gaps of missing data between the reporting sites. Meteorologists must use their best guess as to where an isopleth should be drawn in these areas.

THE MACMILLAN ENCYCLOPEDIA OF WEATHER

Isotherms are one kind of isopleth. They show zones of equal temperature on weather maps.

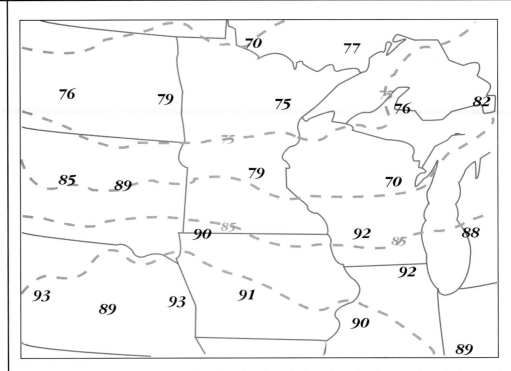

There are several rules for drawing isopleths. One is that an isopleth must either close on itself, forming a loop, or must end at the edge of a map. An isopleth can never split or branch. A second rule is that isopleths are usually drawn with equal intervals. In other words, temperature isopleths will be drawn every ten degrees. Isopleths should also be drawn smoothly and should not connect different points in straight lines.

Meteorologists have distinct names for different isopleths, depending on the data that is being analyzed. ***See also: Weather Maps; Wind; Temperature.***

Different isopleths are given unique names, depending on the data that is analyzed.

Isopleth Name	Data
Isobar	Pressure
Isallobar	Pressure Change
Isotherm	Temperature
Isodrosotherm	Dew Point
Isohyet	Rainfall or Snowfall
Isotach	Wind Speed

Jet Stream

A jet stream is a discontinuous current of air that flows generally from west to east around the globe at high altitudes in the atmosphere. The jet stream is constantly shifting and meandering, forming large bulges that carry air north or south. Jet streams move in response to the movement of large masses of warm and cold air. Jet streams steer weather systems and play a role in their development.

How Jet Streams Are Defined

A jet stream is defined by wind speed and by location. A jet stream has regions of faster and slower winds, and sections where the wind speeds slow so much that the jet stream becomes indistinguishable from the surrounding atmosphere. They are typically found at an altitude of 25 to 35,000 feet and can be hundreds of miles in width.

Winds inside the jet stream can blow anywhere from 50 to 200 mph. Zones of faster wind speeds inside a segment of a jet stream are called jet streaks. A section of the jet stream that curves toward the north, then south again like an upside-down "u" is called a ridge. A trough is a part of the jet stream that curves south, then back north like a right-side-up "u." Jet streams can occasionally split into two separate branches that may later recombine into one jet stream thousands of miles east from where the split occurs.

Formation of Jet Streams

Jet streams form as a result of large-scale temperature differences between tropical and polar regions. Differences in temperature across the globe result in differences in the pressure and density of the air. Because air tends to move from high to low pressure, global temperature and pressure differences result in wind. Jet stream winds occur high in the atmosphere where frictional forces have little effect on the speed of the moving air.

Jet stream winds snake and bend around the globe, rather than moving in a straight line, because the earth is unevenly heated. Oceans warm and cool at a different rate than land areas. Masses of warm and cold air that form as a result of this uneven heating spread across the planet. Jet streams occur where strong temperature differences exist between these masses of moving cold and warm air. Jet streams are more vigorous in winter as colder air spreads farther toward the equator, causing much greater temperature differences.

IMPACT OF JET STREAMS

Jet streams play an important role in the formation, movement, and dissipation of large-scale weather systems. The regions in the jet stream that are most important in the development of weather systems are the areas where the jet stream becomes narrower or wider. Wind speeds tend to increase as air is forced through a narrower segment of a jet stream, much like water flowing faster through a narrow canyon. When wind flows together and increases the amount of air over a given area, the wind is said to converge. Conversely, wind speeds tend to decrease as air moves from a narrower to a wider segment of the jet stream. When wind spreads apart and decreases the amount of air over a given area, the wind is said to diverge.

As the amount of air increases or decreases over a given area because of the convergence or divergence in a jet stream, air pressure at the earth's surface can rise or fall. The large-scale low-pressure areas that spread rain or snow across sections of the country form underneath regions of diverging jet stream flow. The large-scale high-pressure areas that often bring clear, dry weather form underneath regions of converging jet stream flow. As the jet stream bends and moves in response to the movement of cold and warm air masses, the regions of convergent and divergent flow will also move. As these regions move, so will the areas of low and high pressure underneath them.

Areas of divergence and convergence are more numerous and strong in areas where the jet stream shows more pronounced curves and bends. Therefore, the strongest storms occur when jet streams contain large troughs and ridges. This strongly curved jet stream pattern is called meridional flow. By contrast, regions of divergence and convergence are fewer and weaker when jet stream winds blow in a more or less straight line. This relatively flat jet stream pattern is called zonal flow. *See also: Air Mass; Atmosphere; Convergence; Divergence; High Pressure; Low Pressure; Meridional Flow; Ridge; Temperature; Trough; Weather; Wind; Zonal Flow.*

LAKE-EFFECT SNOW

Lake-effect snow is localized, and sometimes heavy, snow that develops downwind of exposed lake water. It is most prevalent south and east of the Great Lakes during early winter; the lake-effect snow that falls downwind of the Great Lakes produces the highest annual snowfall totals anywhere east of the Rocky Mountains.

However, any sufficiently large lake can cause lake-effect snow if conditions

LAKE-EFFECT SNOW

are right. A related phenomenon, called sea-effect snow, occurs when cold air spreads over warmer ocean water.

Lake-effect snow forms when cold air sweeps over an open lake. The air picks up moisture from the lake, which rises and forms clouds that produce snow. Many factors play a role in the development of lake-effect snow. Particularly favorable is a large difference between the temperature of the lake water and the overspreading air. The relatively warm water heats the air in contact with it, which rises into the cold air above. The greater the temperature difference, the faster and higher the air will rise. The rising air eventually cools and forms clouds.

This process may or may not produce snow, depending on how long the cold air is in contact with the water. This depends on wind direction, especially as it relates to the orientation of the lake. Westerly winds blowing over a lake situated along an east-west axis will gather much more moisture than will northerly winds. The length of lake over which air travels is called the fetch. At least fifty miles of fetch is required for significant lake-effect snow.

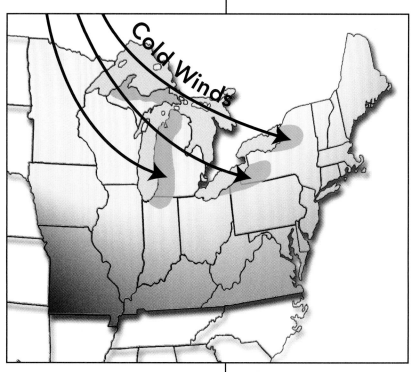

Cold winds blowing across the Great Lakes produce lake-effect snow.

Winds that are aligned (blowing from the same direction) with height can produce heavier lake-effect snow than winds that are sheared (blowing from different directions). Greater wind shear results in disorganized, weaker snow showers, while aligned winds can result in a single, organized, intense band of lake-effect snow.

The change in the elevation of land downwind of a lake also has direct impact on lake-effect snow intensity. If unstable moisture-laden air spreads over steeply-rising terrain, it will rise with even greater vigor. This vertical air motion enhances cloud and snowfall development.

The amount of snow that can fall in lake-effect events can be enormous. One area that experiences these dramatic effects is the Tug Hill Plateau, east of Lake Ontario in northern New York State. The terrain of this area rises over 1,000 feet in a matter of thirty miles. When cold winter winds funnel down the long east-west fetch of Lake Ontario, lake-effect snow can bury this part of the state. Local observers have measured snowfall rates of five inches per hour, accompanied by thunder and lightning. Parts of the Tug Hill Plateau receive an average of 250 inches of snow per winter; by comparison, Chicago averages 37.6 inches of snow annually, and New York City averages 28.4.

Other areas that frequently experience heavy lake-effect snow are parts of western New York State, northeastern Ohio and northwestern Pennsylvania, and the Upper Peninsula of Michigan. Though lake-effect snow occurs over a wide region, individual snowfall events can be extremely localized. Narrow bands of lake-effect snow can form when winds are aligned at different altitudes and blow down the length of a lake. Air rising from the surface of the lake causes surrounding air to flow inward, or converge, to take its place. Narrow swaths of lake-effect snow form along these lines of convergence. These bands of snow can cause blizzard conditions on one side of a city, while the other side remains sunny. *See also: Cloud Formation; Convergence; Lightning; Oceans; Snow; Temperature; Wind; Wind Shear.*

La Niña

La Niña is an area of cooler-than-average ocean water in the tropical eastern Pacific off the coast of South America. La Niña, Spanish for "girl child," is the counterpart of El Niño, which is an area of warmer-than-normal ocean water in this region. Like El Niño, La Niña occurs every few years and can last for up to a year or two.

A strong La Niña affects global weather patterns. When La Niña occurs, certain parts of the planet experience above- or below-normal temperature or precipitation. La Niña, and the related phenomenon El Niño, are currently the subjects of intense scientific research. These events are dramatic examples of how the oceans interact with the atmosphere.

La Niña forms in conjunction with wind and pressure patterns across the tropical Pacific. Normally, an area of relatively cool water and high air pressure exists over the eastern tropical Pacific off of the South American coast. At the other end of the ocean, an area of relatively warm water and low air pressure exists around northern Australia and Indonesia. This warmer water typically extends into the central Pacific. Since air tends to move from higher to lower pressure, winds generally blow from east to west across the Pacific. These winds are known as the trade winds.

During a La Niña event, water over the eastern tropical Pacific becomes even colder and air pressures even higher. Ocean water temperatures may be as much as 7°C below average. Over the central and western Pacific, meanwhile, water temperatures rise and air pressure becomes even lower. As the difference in pressure between the western and eastern Pacific increases, the trade winds increase and spread the cold ocean water away from the South American coast. More cold water rises from deep in the ocean to replace the surface water blown westward by the trade winds. In this way, the area of relatively cold water in the eastern Pacific becomes much larger than normal. Just what causes pressure and temperature patterns across the Pacific to suddenly shift and give rise to a La Niña event is not fully understood at this time.

La Niña can sometimes cause cold and snowy winters over the Midwestern United States.

La Niña affects weather patterns through a complex interaction between oceans and the atmosphere. Colder ocean water causes the air in contact with it to become colder. Colder air contains molecules that, on average, move slower and become more dense. Therefore, air pressure generally rises over regions of colder ocean water. Since the atmosphere is a continuous and flowing layer of air, these La Niña-driven changes in weather over the eastern tropical Pacific eventually lead to changes in weather around the globe.

However, only certain parts of the earth experience noticeable changes in weather conditions as a result of La Niña. In particular, northern Australia and Indonesia have wetter-than-normal weather, while eastern South America becomes drier than normal. In the United States, La Niña produces dry weather in the Southwest, a dry fall season in the central Plains, and a dry winter in the Southeast. The Pacific Northwest, on the other hand, becomes much wetter than average, and winters can be more severe across the Midwest. For reasons not yet fully understood, not every La Niña event produces the same variations in weather.

Scientists have also observed an increased frequency of hurricanes in the Atlantic with strong La Niña conditions in the Pacific. Lower air temperatures over the tropical eastern Pacific cause the temperature difference between middle latitudes and equatorial latitudes to decrease. Since jet stream winds high in the atmosphere blow as a result of horizontal temperature differences, La Niña causes high-level winds over the tropics to weaken. Strong high-level winds shear apart developing Atlantic hurricanes; so when La Niña causes these winds to slacken, Atlantic hurricanes can become more numerous and intense, given other favorable conditions.

While El Niño has gained the most notoriety over the years, some researchers think that La Niña may actually result in greater damage and loss of life, at least in the United States. El Niño causes destructive storms in California but often results in mild winters in the eastern United States and fewer hurricanes in the Atlantic. La Niña, on the other hand, can result in snowy and cold winters in the Midwest and can increase the frequency of hurricanes in the Atlantic. Some researchers also speculate that tornado frequency increases across the Plains states and Midwest during La Niña years.

Meteorologists use the same methods to study La Niña and El Niño. Since both events are changes in ocean water temperatures over the eastern tropical Pacific, scientists use ships and floating buoys to measure water temperatures and take weather observations. Supercomputers run programs designed to simulate the atmosphere and the oceans. These programs, called atmospheric models, can simulate the effects of ocean water temperature changes and lead to a greater understanding of how the oceans and atmosphere are linked. The long-term goals of researchers are to be able to accurately predict La Niña or El Niño conditions far in advance and to fully understand how these events affect weather conditions around the world. *See also: Atmosphere; El Niño; Jet Stream; High Pressure; Low Pressure; Oceans; Precipitation; Temperature; Trade Winds; Weather; Wind.*

LIGHTNING

Lightning is a powerful electrical discharge originating inside a thunderstorm. It occurs after oppositely charged particles of ice build up in separate regions inside a thunderstorm cloud. Lightning can remain in the thunderstorm cloud or strike the ground. Lightning is extremely dangerous when it hits the ground, causing an average of ninety-two deaths and $25 billion in damage in the United States each year.

Lightning is a common occurrence. Around the earth, dozens of lightning bolts hit the ground each second. In the United States, it is estimated that 15–20 million lightning bolts strike the ground each year. Florida suffers the most lightning-related fatalities each year on average, owing to a high thunderstorm frequency and a warm climate, which encourages outdoor activities.

A lightning bolt holds an extreme amount of energy. The power generated by an average lightning strike may be 20,000 amps and can be as high as 100,000 amps. This enormous power heats the air in contact with the lightning bolt to as much as 50,000°F, nearly five times the surface temperature of the sun. This causes a violent shock wave as the air rapidly expands in the extreme heat. The shock wave spreads outward from the lightning bolt and results in what humans know as thunder. Thunder travels at the speed of sound, or 1,087 feet per second. Since one mile equals 5,280 feet, thunder moves about one mile every five seconds. Therefore, when a person sees a lightning flash, he or she can estimate the distance to the lightning bolt by counting the seconds it takes for thunder to reach his or her ears and dividing by five.

Lightning occurs when a significant difference in charge builds up between upper and lower portions of a cumulonimbus, or thunderstorm cloud. An atom becomes charged, either positively or negatively, when it gains or loses electrons. In a cumulonimbus cloud, tiny ice crystals, larger ice pellets called graupel, and hail become charged when they collide with one another. The tiny, lighter ice crystals become positively charged and are blown to the top of the cumulonimbus cloud by strong updrafts. The larger, heavier graupel and

A time-lapse photo shows many bolts of lightning striking the ground during a thunderstorm.

Lightning is extremely dangerous. It can cause fires in houses and buildings, knock out electrical power, ignite forest fires, and cause serious injury or death. There are several basic lightning-related safety tips. When lightning and thunder threaten, the best place to be is indoors away from windows. Avoid telephones, unless they are cordless, and avoid running water. If caught outside, stay clear of tall objects and trees. Lightning is naturally attracted to tall objects. Many deaths and injuries have resulted when people have taken shelter from a thunderstorm underneath a tree, only to have the tree be struck by lightning. When outdoors and away from a building, a car is a good place to seek refuge from lightning.

hailstones become negatively charged and remain mostly in the middle and lower parts of the thunderstorm cloud. The cumulonimbus cloud, therefore, becomes oppositely charged between upper and lower levels of the cloud. An actual lightning discharge occurs when this difference in charge reaches a critical level. It is thought to begin in the lower parts of the cumulonimbus cloud, involving a separate, smaller region of positive charge in the area where rain and hail fall out of the thunderstorm's base. The exact process whereby a lightning bolt begins is still not fully understood.

Once the initial lightning discharge occurs, there is an extremely rapid progression of events that leads to the visible lightning bolt. These events occur so fast that the human eye is unable to see that the lightning bolt is anything but a single flash of light. The first stage in a lightning bolt is called the stepped leader. The stepped leader is a stream of invisible, negatively charged particles that shoots down from the base of the cloud. It proceeds toward the earth in steps, or segments about 160 feet long, at a speed of around 270,000 miles per hour. As the stepped leader nears the ground, streams of positively charged particles shoot upward from trees, buildings, or any other object in contact with the earth. These upward-moving streams of positive energy are faintly visible and have occasionally even been photographed. Once a positive stream emanating from the ground meets up with a stepped leader shooting down from the cloud, an electrical connection is established between cloud and ground.

The second stage in a lightning bolt occurs when this cloud-to-ground connection is made. Positively charged particles then stream upward from the ground to the cloud in what is known as the return stroke. This results in an extremely bright flash of light along the path of electrical current. Even though lightning is an electrical process that begins in the cloud and proceeds to the ground, the visible lightning bolt is actually a stream of energy that starts at the ground and proceeds to the cloud.

Finally, after the initial return stroke, another stream of negatively charged particles can race down from the cloud to the ground. This is called the dart leader. It is followed by another visible flash of positively charged particles from the ground to the cloud. Each lightning bolt can consist of several dart leaders, resulting in multiple lightning flashes. Multiple dart leaders and return strokes over a period of seconds result in a flickering effect.

There are several different kinds of lightning-related phenomena. One is so-called heat lightning. Heat lightning occurs at night when a lightning flash is so far away that thunder cannot be heard. Since someone observing the distant lightning hears no thunder and experiences no rain from the thunderstorm cloud, the lightning flash is mistakenly attributed to the release of energy from heat in the atmosphere.

Another lightning phenomenon is ball lightning. This kind of lightning is rumored to exist through stories and eyewitness events, but scientists have as yet been unable to collect much evidence or data. While photographs of ordinary lightning strokes are common, photographs of ball lightning are extremely rare and are themselves of dubious authenticity.

Two other kinds of lightning-related phenomena, called sprites and blue jets, have only recently been photographically documented. These luminous discharges occur out of the tops of thunderstorm clouds, reaching as high as twenty to fifty miles in the upper atmosphere. Since they occur at such high altitudes and are much less bright than ordinary lightning bolts, they have rarely been seen from the earth. Spectacular photographs of sprites and blue jets, however, have been taken from the space shuttle and from high-flying jet planes. *See also: Air; Benjamin Franklin; Graupel; Hail; Rain; Temperature; Thunderstorms.*

LOCAL WINDS

Local winds are winds that recur in certain locations under similar circumstances. People who experience these winds year after year have named many of them. While most of these winds are no more than meteorological curiosities, some of these winds occasionally become strong enough to cause damage.

One kind of local wind circulation that most people are familiar with is the sea breeze. This kind of wind occurs along the ocean shoreline and results from the difference in heating and cooling rates between ocean and land. During the day, air in contact with sun-heated land becomes much warmer than air in contact with adjacent ocean water. The heated air over the land will expand and become less dense, resulting in lower atmospheric

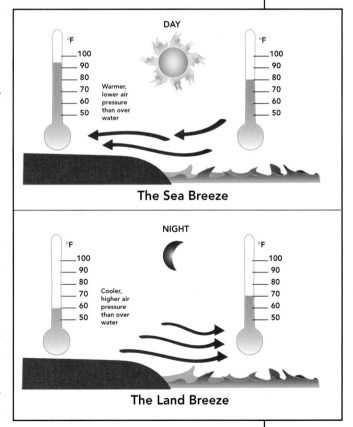

Sea and land breezes occur as the result of temperature differences between land and ocean.

Rugged mountain ranges often help to produce certain kinds of local winds.

pressure than the air over the adjacent ocean. Since wind blows from higher to lower pressure, the daytime heating of the land will result in an onshore-directed wind called the sea breeze. At night, the temperature of air in contact with the rapidly cooling land will become much lower than the temperature of air in contact with the slowly cooling water. This will result in relatively higher air pressure over land compared with that over water. The wind will shift as a result, blowing from land to water in what is known as a land breeze. Similar winds also occur along the shores of large lakes.

Mountain and valley breezes occur through a similar process. During the day, the sun heats the sides of a mountain or ridge. This will cause the air in contact with the heated mountainside to expand and rise up the sides of the mountain, resulting in a valley breeze. At night, the opposite occurs, when the mountainsides cool. Cooler air becomes more dense and flows down the mountain slopes into the valley, in what is known as a mountain breeze.

Another example of a local wind is the fall wind, named not after the season but because this wind falls down mountainsides. It differs from the mountain breeze in its cause. Fall winds blow when large weather systems cause cold, dense air to gather over mountains and plateaus. Like water rushing down a hill, this cold, dense air rushes down from higher to lower elevations. A fall wind that blows down the Rhône valley in France has been named the Mistral, while another fall wind that blows down from the mountains of Bosnia and into the Adriatic is called the Bora.

Foehn winds also blow from higher to lower elevations, but they are distinguished from fall winds in that foehns are typically warming winds. The word "foehn" originated in Europe as a term used to describe warming winds blowing down from the Alps, but it has since spread to similar winds all over the world. In the United States, a type of foehn that blows down from the Rocky Mountains onto the high plains from Montana to Colorado is called the Chinook. It is known for bringing rapid thaws and snowmelt after winter cold snaps. Another type of foehn, blowing in southern California, is called the Santa Ana. This wind blows down from the coastal mountain ranges into the highly populated areas of the southern California coast. At their windiest and driest, Santa Anas can contribute to devastating wild fires.

Other local winds around the world include the Simoom, which blows from Egypt into the Middle East and is known for extremely hot, dry, and dusty air. The Kona is a rain-bearing wind that blows from the southwest in Hawaii. In Australia, the Brickfielder is a hot and dusty wind that blows from the interior deserts onto the coastline. And near Santa Barbara, California, the Sundowner is a rare, hot wind that blows down the mountains in the evening, causing temperatures to rise above 100° in a matter of minutes. *See also: Air; Oceans; Ridge; Temperature; Wind.*

LOW LEVEL JET

A low level jet is a region of fast-flowing air in low levels of the atmosphere, generally located anywhere from several hundred to 5,000 feet in altitude and extending several hundred miles in length. Wind speeds in low-level jets are generally from 30 to 60 mph. By contrast, the upper-level jet stream (often just referred to as the jet stream) is a discontinuous river of air that stretches around the planet at altitudes of 25,000 to 35,000 feet and blows at speeds from 60 to 120 mph. Low-level jet streams often form in association with areas of low pressure and contribute to regions of heavy rainfall or snowfall.

One example of a low-level jet occurs over the central United States at night during spring and summer. During this part of the year, strong southerly winds often blow from the Gulf of Mexico northward into the Plains or Midwest. During the day, eddies and updrafts generated by the sun's heating mix with the southerly flow. This turbulence tends to disrupt and slow the southerly winds. At night, however, the turbulence dissipates and wind speeds increase. The low-level jet develops in this accelerated flow several hundred to several thousand feet above the ground where there is no friction to slow the air. The southerly low-level jet often transports abundant tropical moisture northward into the central United States. This humidity frequently flows into large clusters of thunderstorms that develop during the night over the Midwest and Plains states. Weather forecasters look at wind data over the central United States to see whether a low level jet has formed and where it is directed. Knowing this enables them to better predict where nighttime thunderstorms will occur.

Low-level jets also occur in conjunction with strong areas of low pressure. Meteorologists studying wintertime low-pressure areas along the East Coast of the United States found that a low-level jet usually forms just ahead of the advancing area of low pressure. This low-level jet, called a cold conveyor belt, flows from east to west and transports cold, moist air into the storm system. This cold, moist air, originating over the Atlantic Ocean, enhances precipitation rates and often results in heavy snow north of the center of low pressure. ***See also: Air; Atmosphere; Jet Stream; Low Pressure; Oceans; Precipitation; Thunderstorms; Turbulence; Wind.***

LOW PRESSURE

A low-pressure area, or "low," is a large atmospheric circulation, the center of which has a lower atmospheric pressure than all surrounding areas. Winds spiral counterclockwise and inward toward low-pressure systems. Depending on its strength, a low can be accompanied by unsettled weather, including widespread cloudiness and precipitation.

Certain patterns of wind favor the development and decay of low-pressure areas. These wind patterns, known as converging and diverging winds by meteorologists, occur at the surface of the earth as well as at high elevations in the atmosphere. A diverging wind spreads air apart, decreasing the total mass of air over an area. A converging wind brings air together, increasing the mass of air over an area.

Low-pressure areas generally form near ground level, underneath areas of diverging wind at high elevations in the atmosphere—20,000 to 30,000 feet. The more air removed by diverging winds from high in the atmosphere, the less air remains to weigh down on the surface of the earth; and consequently, the lower the air pressure becomes. Winds frequently diverge and spread apart when they slow down. These decelerating winds, and regions of divergence, usually occur ahead of advancing zones of cold air at high altitudes in the atmosphere.

As air is removed aloft, other air rises up from the earth's surface to take its place. This rising air, in turn, is replaced by winds that blow together, or converge, horizontally near ground level toward the center of the low-pressure area. As long as more air diverges aloft than converges at the surface, air pressure will fall and the low-pressure area will strengthen, or deepen.

Since winds in the upper levels of the atmosphere are constantly moving, areas of upper-level divergence are constantly moving as well. As the zone of diverging air aloft moves along, surface air pressure lowers beneath it. After the diverging air passes an area, the air pressures rises. In this way, an area of low pressure moves as surface air pressures lower ahead of, and rise behind, upper-level divergence. If the area of upper-level divergence weakens—or if the amount of air converging at the earth's surface towards the center of the low becomes larger than the amount of air diverging at high altitudes—the area of low pressure will weaken. When a low-pressure area weakens, meteorologists say it "fills," referring to the increasing quantity of air molecules filling the center of the low-pressure area and causing air pressure to rise.

The rising air in low-pressure areas is responsible for clouds and precipitation that frequently accompany lows. Clouds form when rising air carries water vapor upward into the atmosphere. As the water vapor cools, it changes from gaseous molecules to liquid droplets, forming clouds. The higher the air rises into the atmosphere and the more water vapor in the air, the larger the clouds become until eventually they produce rain or snow.

Low-pressure areas that develop outside of the tropics are called extratropical lows. They are characterized by colder air near the center of the system and by warm and cold fronts that connect to the center of low pressure. The fronts mark the leading edges of warmer and colder air circulating around the center of the low. Low-pressure areas that develop in the tropics are called tropical lows. These systems, which include tropical depressions, tropical storms, and hurricanes, are characterized by warmer air near the center of the system and are not accompanied by any fronts. Unlike extratropical lows, tropical lows require warm ocean water to grow and intensify. They dissipate when they move over land or over cold ocean water. *See also: Air; Atmosphere; Cloud Formation; Cold Fronts; Divergence; Oceans; Precipitation; Rain; Snow; Water Vapor; Wind.*

Low-pressure areas often contribute to the formation of clouds and rain or snow.

Meridional Flow

The term "meridional flow" describes the flow of the jet stream in upper levels of the atmosphere. The jet stream is a discontinuous current of fast-moving air at high altitudes, typically 20,000 to 30,000 feet in the atmosphere. It generally flows from west to east but can meander north and south in curves and bends that extend for thousands of miles. In general, a highly curved jet stream flow is said to be meridional, while a straighter, west to east flow is said to be zonal. When the upper-level air flow is meridional, areas of low and high pressure at the earth's surface are likely to be much stronger.

The jet stream can be likened to the flow of water in a river. Sometimes rivers flow nearly straight, while other times they snake and curve in wide loops back and forth across the land. Meridional flow is similar to a highly curved and undulating river. The direction and speed of jet stream winds is determined by the movement of large masses of cold and warm air around the earth. Jet stream winds blow in between these air masses, in the regions of greatest horizontal temperature contrast. The greater the temperature contrast, the stronger the wind. The direction of air flow is determined by the position of the air masses relative to one another. Jet stream winds blow southward around a large cold air mass, then back northward around a large warm air mass.

Meridional flow is associated with strong areas of low and high pressure at the earth's surface. These low- and high-pressure areas form underneath regions of diverging and converging winds, respectively. Diverging and converging wind flow removes or adds air over a region, thereby affecting air pressure at the earth's surface. Strong divergence, for example, can remove air from over a region and cause an area of low pressure to form or strengthen near ground level. When the upper-level air flow is meridional, there are more pronounced regions of divergence and convergence than if the air flow simply moves from west to east. Areas of low- and high-pressure, therefore, become much stronger on average. *See also: Air Mass; Atmosphere; Convergence; Divergence; High Pressure; Jet Stream; Low Pressure; Temperature; Wind.*

Mesocyclone

A mesocyclone is a large, rotating column of air inside a thunderstorm. Meteorologists will probe a thunderstorm with Doppler radar to determine

whether or not it contains a mesocyclone. Atmospheric scientists have long observed that tornadoes often develop within the circulation of a thunderstorm's mesocyclone. Considerable effort is being spent to determine the chain of events that leads to the formation of a tornado from a mesocyclone circulation.

HOW MESOCYCLONES FORM

A mesocyclone forms in the updraft of a thunderstorm. An updraft is a current of air that rises vertically into the thunderstorm cloud, transporting warm, moist air from the earth's surface upward into the storm. An updraft can extend from the base of the thunderstorm cloud upward to nearly the top of the cloud, tens of thousands of feet in altitude. Long-lived, intense thunderstorms called supercells contain an updraft that rotates in a counterclockwise direction. This rotating updraft, around a half mile to a mile wide, is the mesocyclone. Air spirals upward through the cloud in a corkscrew fashion inside a mesocylone.

All thunderstorms contain at least one updraft, but only a small fraction contain a mesocyclone. A mesocyclone forms when significant wind shear is present in the atmosphere. Wind shear is the change of wind speed and direction with increasing height. Scientists believe an updraft gains rotation as the result of wind shear present in the atmosphere before the thunderstorm even develops. This wind shear causes air in low levels of the atmosphere to roll about a horizontal axis. When the updraft in a building thunderstorm rises through the atmosphere, it tilts one of these horizontal rolls into a vertical position. As the updraft rises into this spiraling column of air, it begins to rotate and becomes a mesocyclone.

This photograph shows the part of a thunderstorm cloud that contains a mesocyclone.

Mesocyclones and the Formation of Tornadoes

Tornadoes develop from mesocyclones in a complex process not yet understood. The spinning circulation of a mesocyclone is not enough, by itself, to produce a tornado, and recent research indicates that less than 50 percent of all mesocyclones produce tornadoes.

In general, a tornado forms when the rate of spin increases in a small area within the larger mesocyclone circulation. One theory on tornado formation proposes that the mesocyclone intensifies as a result of a complex interaction with the thunderstorm's downdraft. A downdraft is a descending current of air inside a thunderstorm cell that forms when heavy rain and hail drags air downward towards the earth. Recent experiments in the field, where scientists observed actual tornadoes and mesocyclones from close range, even revealed that some strong tornadoes can form outside of and separate from a thunderstorm's mesocyclone. Researchers are using the data collected in these field experiments to run computer simulations of thunderstorms in an attempt to shed more light on how mesocyclones form and the nature of their relation to tornadoes. *See also: Air; Doppler Radar; Hail; Low Pressure; Rain; Thunderstorms; Tornadoes; Wind; Wind Shear.*

Mesoscale Convective Complex

A mesoscale convective complex (MCC) is a large conglomeration of thunderstorms that typically develops and reaches peak intensity during the nighttime hours. An MCC can cover tens of thousands of square miles and last for twelve hours as it moves along for hundreds of miles, often tracking over several states. MCCs occur most often over the central United States during the spring and summer, and can produce heavy rain, strong winds and hail. While MCCs can sometimes cause flooding, the rain they produce is vitally important to crops in the Plains and Midwest.

Characteristics of MCCs

MCCs differ from ordinary thunderstorms in the way they develop, their size, and the time of day in which they occur. Most thunderstorms develop after the sun warms the air during the day, then dissipate after sunset. MCCs, however, are primarily a nighttime phenomena associated with a meteorological feature called the low-level jet. While low level jets can occur in a variety of meteorological situations, the kind that contributes to MCC formation is a region of strong southerly winds at low levels in the atmosphere in the central United States. It often forms during spring and summer when the general wind flow is from the south. During the day, turbulent air currents and eddies generated by the sun's heating mix with and slow the southerly wind flow. After the sun goes down, however, the turbulent interference ceases and the southerly winds can strengthen to 30–60 mph in a current hundreds of miles long, situated from several hundred to several thousand feet above ground level. The low level jet transports rich tropical moisture and warmth northward from the southern U.S. and Gulf of Mexico.

An MCC will often form as the northward-directed low-level jet intercepts a front lying from east to west across the central United States. As the warm moist air in the low-level jet collides with cooler, drier, and denser air along the front, it rises into the atmosphere. The rising motion carries moisture high into the atmosphere, where it condenses into cloud droplets. The strong southerly winds in the jet pump vast quantities of moisture and heat upwards into the developing clouds, causing them to quickly become thunderstorm clouds. As these thunderstorms strengthen and merge together, an MCC develops. An MCC can grow to cover the better part of a state, producing torrential rain, hail, and sometimes damaging winds. MCCs generally move from west to east, advancing from state to state through the nighttime hours. As the sun rises and heats the land the next day, turbulent eddies will once again disrupt the low-level jet. As its supply of moisture gets cut off, the MCC itself will dissipate.

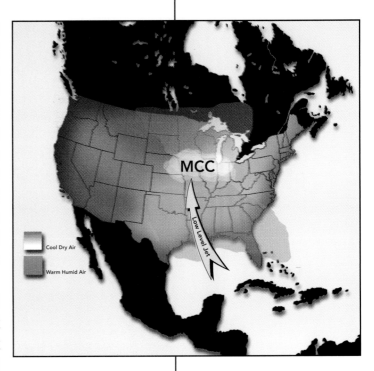

MCCs often form in the central United States when strong southerly winds carry moisture northward from the Gulf of Mexico.

Meteorologists have observed that MCCs can go through two to three cycles of nighttime growth and daytime dissipation. As the MCC flares up each night it produces an area of low pressure and a wind circulation at mid-levels of the atmosphere, around 17,000 feet up. Even though the clouds and rainfall associated with the MCC dissipate after sunrise, the low-pressure area and circulation often persist through the day. The next night, this small-scale weather system can become the focus for the re-development of the MCC.

TRACKING MCCs

Forecasters track MCCs using satellite and radar. During the darkness of night, clouds are invisible to visual satellite photography. But satellites can sense the presence of clouds by detecting the infrared radiation that they emit. Infrared, or

IR, satellite photographs, show the temperatures of the cloud tops. In general, the higher the cloud the colder the cloud's temperature. Thunderstorm clouds associated with MCCs often tower over 50,000 feet in the atmosphere, with cloud top temperatures as low as -50°C. Forecasters easily identify an MCC on IR satellite photographs as a very large, circular area of extremely cold clouds over the central United states. Radar helps meteorologists track the MCC at ground level by showing where the heaviest rain is occurring and the direction the MCC is moving. *See also: Air; Atmosphere; Hail; Low Level Jet; Low Pressure; Rain; Radar; Radiation; Satellite; Temperature; Thunderstorms; Wind.*

METEOROLOGY

Storm clouds in the Gallatin Mountains of Southern Montana.

Meteorology is the study of the earth's atmosphere. It encompasses the observation and forecasting of weather over short time periods, as well as the study of weather over long time periods and the interaction between atmosphere and ocean. Meteorologists have a variety of tools at their disposal to help them in their task, including leading-edge computer, electronic, and space technology.

ASPECTS OF METEOROLOGY

There are a few ways to subdivide the science of meteorology. One way is to consider the time period over which the atmosphere is studied. Weather forecasters, for example, observe and predict the weather over a period of hours, days, and sometimes weeks. Climatologists, on the other hand, study the long-term average of weather conditions in a given region over a period of years. The atmosphere can be studied on ever longer time scales; paleoclimatologists examine evidence showing how the earth's atmosphere existed for thousands or even millions of years.

Meteorology can also be considered relative to how large a section of the atmosphere the meteorologist examines. The size of this section is called the "scale" of the atmosphere, and there are four main scales in which meteorologists work. Microscale studies deal with atmospheric processes occurring in an area ranging from yards to miles in diameter and on time scales from seconds to minutes. Turbulence, such as that which affects airplanes, is a microscale phenomena. Mesoscale meteorology is the study of atmospheric events in an area ranging from tens to a hundred or so miles in diameter, and on time scales of minutes to hours. Thunderstorms are one example of a mesoscale phenomena. On the synoptic scale, meteorologists view weather systems that extend from hundreds to thousands of miles across and last for days. Warm fronts, cold fronts, and most areas of low and high pressure are synoptic scale weather systems. Finally, planetary scale meteorology considers the big picture of air, heat, and moisture transfer across the globe over a period of weeks, months, and years.

Another way to subdivide meteorology is in terms of the kind of atmospheric phenomena being studied. Atmospheric chemists, for instance, examine the molecular properties of air. They may study pollution, or the processes which cause the protective ozone layer to break down high in the atmosphere. Hydrologists, meanwhile, study water as it cycles from the atmosphere to the earth and back again. Hydrologists are interested in streams and rivers, and how they flood during periods of excessive precipitation. Tropical meteorologists study the weather only in the tropics, and are especially interested in hurricanes. Dynamic meteorologists study the physical laws that govern the motions of air in the atmosphere, and often use their skills to help develop computer software that simulates the behavior of the atmosphere.

Tools Used In Meteorology

To assist in their investigations of the atmosphere, meteorologists employ a wide variety of tools. Radar is one such tool, allowing weather forecasters to track areas of rain and snow over an area of several hundred miles in diameter. Radar sends out microwave energy that bounces off raindrops, snowflakes, and other small airborne objects. As this reflected energy is received back at the radar, it is electronically processed to determine the location, intensity, and movement of areas of rain and snow. Nearly 150 radar sites operate continuously across the United States. Computer software can combine the images from all the radar sites in order to create a large radar picture that spans the continent and permits forecasters to fully see precipitation patterns associated with large-scale weather systems.

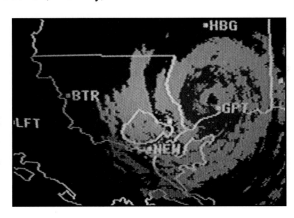

Radar is an indispensible tool for meteorologists. This radar image shows a hurricane moving onshore in southern Mississippi.

Satellites help meteorologists monitor weather conditions around the world.

Satellite imagery also allows meteorologists to see weather patterns that extend over thousands of miles across the earth. Orbiting satellites take frequent photographs of the atmosphere and send the information back to earth where it is processed into images that are displayed at computer workstations. Satellites can take black and white photographs of cloud patterns, similar to those taken by an ordinary camera. They can also take photographs of infrared energy emitted by the atmosphere and of the water vapor content in the atmosphere. This allows meteorologists to continue to track cloud patterns at night, when darkness would otherwise prevent satellite photography. Satellites are useful not only for tracking large-scale weather systems, but also for sensing the weather conditions over remote and uninhabited regions such as oceans.

Weather observations by humans or automated sensors provide real-time data on current weather conditions around the globe. Thousands of observing sites are located across the United States, many of them situated at airports. These weather stations take hourly readings of temperature, moisture, wind, air pressure, weather, and cloud cover. This data not only helps weather forecasters know what the weather is doing currently, but serves as input for extremely complex computer programs that simulate weather patterns over days. This software runs on large mainframe computers, called supercomputers, and predicts weather patterns across large parts of the earth for days in advance. Supercomputers can quickly and efficiently process the enormous amount of complex mathematical calculations necessary to predict the development and movement of weather systems. Computers are the backbone of meteorology, assisting in everything from the processing and storing of radar and satellite data to the creation and display of television weather graphics.

The most important meteorological tool, however, remains the human brain. Meteorologists develop hypotheses about the atmosphere and how it works, and devise ways to test these hypothesis. The mind of a meteorologist contains conceptual models of atmospheric processes, allowing the person to consider weather conditions in light of their understanding of the physical laws which govern the atmosphere, and with respect to the person's experience in observing how the atmosphere behaves. Meteorologists evaluate the output from computers that predict the weather, and consider how the computer's output might be wrong as a result of known inadequacies of the computer software, or of the quality of the data that the computer used as input. Meteorologists make split-second decisions to issue life-saving warnings for tornadoes or flash floods. While computers are becoming increasingly important for meteorologists, the human capacity for experience, intuition, and decision-making remains the most valuable element in meteorology.

Careers in Meteorology

Meteorology is a rewarding occupation involving the opportunity to work with state-of-the-art technology to study a vast, chaotic, violent, and fascinating natural system. Since meteorology is a science, a successful education involves gaining a proficiency in scientific, mathematical, and computer-related knowledge and skills.

Meteorologists work with advanced computer technology to analyze and forecast the weather.

The most important requirement for a career in meteorology is a passion for some aspect of the atmosphere. Whether it be snowstorms, hurricanes, tornadoes, climate, the greenhouse effect, or the physical laws that underlay these phenomena, an innate sense of excitement and curiosity about the weather is vital for a satisfying career in meteorology.

While most high schools do not offer specialized classes in meteorology, there are three areas of study which will better prepare a student for an undergraduate major in meteorology: math, science, and computers. A high school student interested in a meteorology major will also find a wealth of information on the Internet. Output from the very same computer models that professional meteorologists use is available on dozens of Web sites, mostly associated with universities that offer programs in atmospheric science. These computer models simulate and predict the weather for days in advance. Students can also find weather discussions written by lead forecasters from dozens of National Weather Service (NWS) offices around the country several times per day. Most NWS offices now have Web sites that post these weather discussions, along with most other weather-related products they provide. Radar and satellite images, along with raw weather observations, are also freely available. The Internet offers a wonderful opportunity to gain an early familiarity with some of the tools that professional meteorologists use every day.

There are dozens of colleges and universities that offer degrees in meteorology. The definitive listing for the prospective meteorology major can be found in the book, *Curricula in the Atmospheric, Oceanic, Hydrologic, and Related Sciences.* This book contains course information and contact addresses for meteorology and related departments in universities across North America. It is available from the American Meteorological Society (AMS), 45 Beacon St., Boston, MA, 02108-3693 and is also available at the AMS Web site (http://www.ametsoc.org/AMS).

Once in college, undergraduate meteorology involves a concentration in mathematics, physics, and computer science. A meteorology major is a relatively difficult course of study owing to the amount of classwork involving advanced calculus and physics. Many universities do not even offer actual meteorology classes to meteorology majors until they reach their second or third year. Students successfully completing undergraduate work will sometimes continue into graduate level studies, earning a master's or Ph.D. in meteorology and typically specializing in the area that interests them the most. These advanced degrees are becoming increasingly important for gainful employment as a professional meteorologist.
See also: Air; Atmosphere; Climate; Cold Front; Floods; Greenhouse Effect; High Pressure; Low Pressure; National Weather Service; Ozone; Oceans; Pollution; Precipitation; Radar; Rain; Satellite; Snow; Thunderstorms; Tornadoes; Turbulence; Warm Front; Water Vapor; Weather; Wind.

MICROCLIMATE

A microclimate is a small-scale region in which average weather conditions are measurably different from a larger, surrounding region. These differences can be in temperature, precipitation, wind, or cloud cover, and often result when natural or man-made land features alter local weather patterns. A microclimate can

Microclimate

Microclimates often exist on and around mountains, owing in part to their rugged and varied terrain.

allow certain kinds of plant or animal life to thrive in an area where they are normally not found.

One of the most frequent causes of a microclimate is a significant variation in elevation over a small area. Mountains often harbor microclimates that exist for a couple of reasons. One is that the average air temperature decreases with increasing height. Tops of mountains typically have a much colder climate than the surrounding lower elevations. Certain species of plants and animals that are normally found in colder regions much closer to the poles can exist on the tops of mountains in otherwise warmer climates.

Mountains also create microclimates by altering local patterns of wind and precipitation. Air that blows into a mountain range is forced upwards, cooling and expanding as it rises. As the rising air cools, water vapor mixed in with the air will condense into liquid cloud droplets. When moisture-laden wind continuously blows into and rises up a mountain range, the resulting clouds frequently become large enough to produce rain or snow. As the wind descends the opposite side of the mountain, the moisture quickly evaporates, causing precipitation and cloudiness to dissipate. Mountain ranges, therefore, often have wet microclimates on slopes that face towards the prevailing wind, and dry microclimates on slopes that face away from the prevailing wind.

Microclimates also exist along the shorelines of oceans and large lakes. Since water cools and warms much slower than adjacent land, the temperature range of a coastal region will not usually be as extreme as areas located farther inland. Coastlines will be milder on average than inland locations during winter, and cooler on average than inland locations during summer.

Man-made structures also result in microclimates. Buildings alter the flow of air in such a way as to sometimes produce a windy microclimate. Wind that blows through narrow gaps between buildings will speed up in much the same way that water rushes faster through a narrow canyon. A more significant microclimate that often exists around cities is called the urban heat island effect. Asphalt and concrete in cities absorbs more heat from the sun than vegetation in suburban or rural areas. At night, the city only slowly releases the heat absorbed during the day. The energy released into the air from the heating and cooling of buildings also causes the temperature to rise over cities. During winter the effect is especially noticeable, with cities remaining ten or more degrees warmer than surrounding locations at night. ***See also: Air; Oceans; Precipitation; Rain; Snow; Temperature; Water Vapor; Weather; Wind, Urban Heat Island.***

MIRAGE

A mirage appears when light reflected from an object changes path after passing through air of varying density. This causes the object to appear above or below its actual position. Most people are familiar with the mirages that occur near the earth's surface, such as apparent puddles of water that appear and disappear from a road during a hot day.

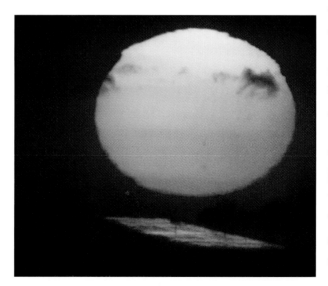

Mirages are the product of sunlight interacting with air of varying intensity.

Near ground level mirages occur in regions where the temperature changes very rapidly with increasing height. Since the density of air molecules is a function of the air temperature, these large vertical temperature differences result in large vertical air density variations.

Mirage

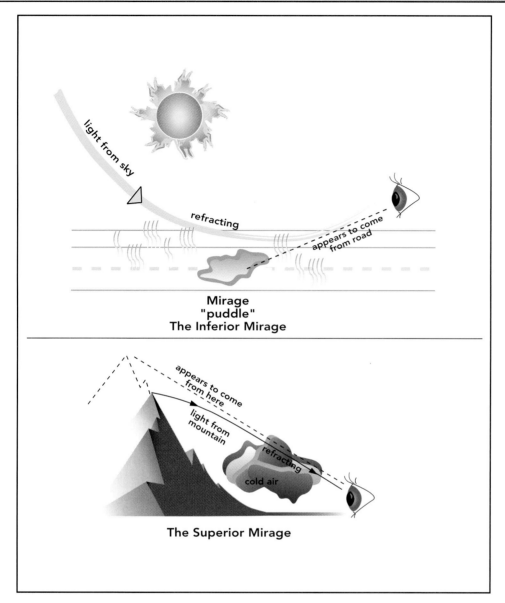

Mirages occur as light bends, or refracts, through air.

As light passes through a denser layer of air it will change course, or refract. To the human eye, the light will therefore appear to originate from above or below its true source. The object from which the light is reflected, therefore, will appear above or below its actual position.

A common type of mirage is called an inferior mirage, because objects appear lower than their actual position. This kind of mirage is responsible for the shimmering puddles of water that seem to appear and disappear on the surface of a road during a hot day. A road surface will become very hot because its dark color and molecular composition absorb, rather than reflect, much of the sun's energy that strikes it. The temperature of the air, therefore, will become very high in a shallow layer just above the road surface then will decrease rapidly several feet up. Light passing through this zone of strongly changing air temperature and density will refract upwards. To the eye it seems as through the light is coming from below the actual position of the object. What appears as puddles of water on the road surface, therefore, is actually refracting light from the sky, trees, cars, or other objects.

The opposite kind of mirage, called a superior mirage, occurs when objects seem to appear above their actual position. Superior mirages happen when very cold air near the earth's surface is overlaid by warmer air. Light originating from objects near the ground refracts downward as it passes through the zone of rapidly decreasing air density. This causes light reflected by objects at ground level to seem to come from above the earth's surface. Owing to presence of very cold air near the ground, superior mirages often occur in the Arctic. **See also: Air; Temperature.**

Monsoon

Parts of the world experience seasonal shifts in wind direction, such that winds blow from one direction during the warm season and from another direction during the cold season. The monsoon is best known as a wet, warm-season wind carrying drenching rains. The term monsoon, however, can also be used to describe the wintertime wind shift that carries dry, cooler air. The word monsoon is derived from the Arabic mawsim, meaning season.

What Causes The Monsoon

The monsoon is most pronounced from the Arabian Peninsula eastward across India and into southeastern Asia. A variety of geographical and climatic influences contribute to its development. In winter, a large, cold high pressure area, or anticyclone, usually covers the interior of Asia. This semi-permanent seasonal weather system is accompanied by cold, dry air. The relatively warm water of the Indian Ocean, however, keeps the air above it at a higher temperature than air over land. Since higher temperatures are associated with lower air pressure, a significant pressure difference develops between the higher air pressure over land and lower air pressure over the ocean. This pressure gradient results in dry, cool offshore winds as air flows from high to low pressure.

During summer, however, the situation changes. As the sun rises higher in the sky, the land becomes warmer. This causes air pressures over land to fall, and the anticyclone over the interior of Asia begins to dissipate. Temperatures over land become much warmer than the adjacent ocean, and the pressure gradient reverses. Air begins to flow from higher pressure over water to lower pressure over land. Since these winds originate over the tropical ocean water, they carry abundant moisture. This contributes to the formation of clouds and the copious rainfall for which the monsoon is known. Tropical weather disturbances

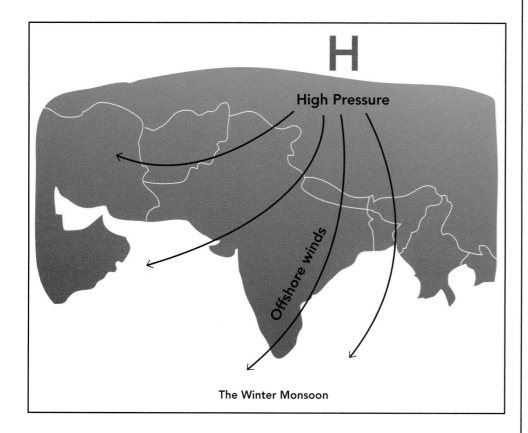

During the winter monsoon over India, dry, cool air flows offshore from a high-pressure area over the Himalaya plateau.

called easterly waves contribute to the heavy rainfall.

Cloud formation acts to strengthen the summertime monsoon. As water vapor condenses into clouds over land, it releases energy into the air. This energy, known as the latent heat of condensation, serves to increases the air temperature surrounding the clouds. The increased air temperature causes surface air pressure to fall, which intensifies the pressure difference between land and ocean. Thus wind flow from ocean to land is reinforced in this positive feedback mechanism.

Effects of the Monsoon

In India, the wet and dry monsoon bring markedly contrasting weather. During the dry monsoon, when winds blow from land to water, rain becomes very infrequent or even ceases completely for months at a time. By the end of the dry monsoon, severe drought conditions and lack of water can bring hardship to humans and animals alike. The arrival of the wet monsoon is a major event and general cause for celebration. It offers cooling relief from weeks of intense heat,

The Macmillan Encyclopedia of Weather

Indians wade through the flooded streets of Bombay during the monsoon season.

and replenishes the water supply as moist, rain-bearing tropical air flows inland.

The wet monsoon, while welcomed, can also cause damage as torrential rain leads to areas of flooding. Rainfall amounts in the summer monsoon over India have been enormous. Cherrapunji, India, located in the hilly northeastern part of the country, recorded 366 inches of monsoon rainfall in just one month. (By comparison, Seattle, Washington, receives an average of slightly less than forty inches of precipitation during an entire year.) **See also: Air; Cloud Formation; Drought; Easterly Wave; High Pressure; Low Pressure; Oceans; Precipitation; Rain; Temperature; Water Vapor; Weather; Wind.**

National Weather Service

The National Weather Service (NWS) is a government agency that issues weather forecasts and warnings to the public twenty-four hours a day. The NWS is one part of a larger organization called the National Oceanic and Atmospheric Administration (NOAA) which exists as part of the U.S. Department of Commerce.

The NWS is comprised of over 120 local offices spread across the country. These offices are staffed by professional meteorologists who monitor weather conditions over a specific localized region or forecast area. Each forecast area is further divided into forecast zones. These forecast zones are arranged in such a way as to reflect the variations in climate that occur as the result of natural changes in geography. Coastal areas, for example, might constitute one forecast zone, while a mountain range farther inland will be assigned a separate forecast zone. NWS meteorologists issue advisories and warnings for severe weather by individual zones, or by counties depending on the kind of advisory being issued. A tornado warning, for example, may be issued for one or two specific counties within a forecast zone. A winter storm watch, on the other hand, would likely be issued for several zones within the forecast region.

NWS meteorologists prepare a variety of weather forecasts. These include short-term forecasts, which provide information on weather conditions expected over a local area for the next six hours or less. A zone forecast, on the other hand, will carry standard weather forecast information divided up into forecast periods such as "today," "tonight," and "Wednesday" for the next two days. The National Weather Service also issues special forecasts for airports, meant to be used by the professional aviation industry. Marine forecasts are prepared for public and commercial boaters in lake and ocean waters, and include information such as predicted winds, wave heights, and times of high tide. Another kind of forecast offered by the NWS is a river forecast. These specialized forecasts indicate the current and predicted water levels in rivers and streams, and are particularly important when heavy rain raises the threat for river flooding.

Weather Advisories

The most important job of a National Weather Service meteorologist is to issue advisories that alert the public in times of severe and potentially life-threatening weather. The National Weather Service is the only authorized source for severe weather advisories in the United States.

There are two main kinds of severe weather advisory: a watch and a warning. A severe weather watch means conditions are favorable for the occurrence of a particular type of severe weather in and near the watch area. A winter storm watch, for instance, indicates that heavy snow is possible over the watch area in the near future. The heavy snow may or may not develop, but people need to know that the possibility exists. Watches are issued for large areas, on the order of hundreds of miles wide. A severe weather warning, on the other hand, is much more serious. It means that a particular kind of severe weather has actually been sighted or indicated by radar. A tornado warning, for example, warns the public that a tornado is actually occurring and heading towards the warned area.

Warnings are localized and usually issued for one or more counties. People in the warned counties should take immediate action to protect life and property. Other kinds of watches and warnings are issued for high winds, floods, hurricanes and tropical storms, blizzards, and severe thunderstorms.

National Weather Service forecasts, watches, and warnings are communicated through a variety of channels including radio, television, and the Internet. A particularly effective means of communication, however, is NOAA Weather Radio. The National Weather Service broadcasts weather forecasts and advisories twenty-four hours a day on special radio frequencies not receivable by standard AM or FM radios. Weather radios, which are small and inexpensive, can be bought wherever standard radios are sold. A valuable feature of most of these radios is the tone alert system. Whenever the NWS issues a severe weather warning, the NOAA Weather Radio will sound an alarm. In this way a person is alerted to threatening weather conditions even if they are not tuned in to standard radio or television.

ADVANCES IN THE NATIONAL WEATHER SERVICE

The National Weather Service continues to modernize and evolve. In the 1990s a fleet of nearly 150 Doppler radars was deployed across the United States. Doppler radar has the capability not only to sense the location and intensity of rain or snow, but also to sense wind flow inside clouds. This allows meteorologists to better see the circulations that sometimes precede the development of tornadoes inside severe thunderstorms. The NWS also uses specialized computer hardware and software that assists forecasters in visualizing weather patterns and manipulating meteorological data. As computer power increases through the twenty-first century, National Weather Service forecasts will likely become more and more accurate. *See also: Climate; Doppler Radar; Floods; Radar; Rain; Snow; Thunderstorms; Tornadoes; Weather; Weather Forecasting; Wind; Winter Storm.*

Nor'easter

Nor'easters are strong areas low pressure, called extratropical cyclones by meteorologists, that move northward along the United States Eastern Seaboard. Nor'easters derive their name from the strong, northeasterly gales that blow in advance of the storm. They mainly occur during fall and winter, and can produce heavy rain or snow, strong winds, and heavy surf and tidal flooding along coastal

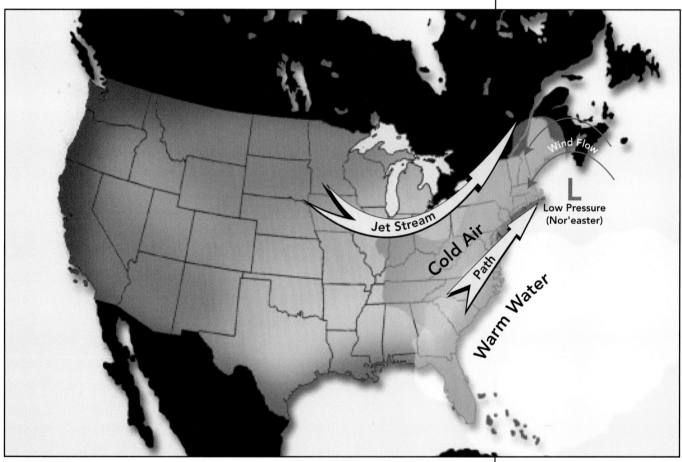

Nor'easters form along the East Coast of the United States where cold air over land meets up with warmer offshore water.

locations. At their strongest, nor'easters can produce as much damage as a minimal hurricane. These large storms affect the most densely populated part of the United States, disrupting travel and affecting millions of people.

Nor'easters form during the colder months of the year when strong jet stream winds move southward from Canada into the United States. The jet stream is a discontinuous current of fast-moving air high in the atmosphere. These fast winds are the result of strong horizontal temperature differences in the atmosphere. The jet stream migrates towards the poles and weakens during summer when temperature contrasts are weak, then moves back south and strengthens during the winter as temperature contrasts increase. It plays an important role in the formation and movement of storms such as nor'easters.

Like many extratropical cyclones, nor'easters form in association with strong jet stream disturbances. A jet stream disturbance consists of a large mass of cold air at high elevations in the atmosphere. As the temperature contrast increases between the cold air aloft and warmer air ahead of it, jet stream winds begin to blow faster. These winds blow in a "u"-shaped pattern, first diving south, then east and back north around the leading edge of the advancing cold air aloft. This "u"-shaped pattern of wind is called a trough by meteorologists. Frequently, winds will spread apart and diverge ahead of a trough. Diverging winds remove air molecules from over a region, causing the total amount of air weighing down on the earth's surface to decrease. Surface air pressure therefore drops underneath areas of diverging jet stream winds. This falling air pressure results in the development of a low-pressure area.

Nor'easters can develop when diverging jet stream winds spread over the East Coast of the United States from fall through early spring. During this time of year, the temperature contrast is greatest between the relatively warm Gulf Stream waters that flow northeastward off the East Coast, and cold air over interior sections. Since jet stream winds tend to increase in regions of strong horizontal temperature contrast, jet stream disturbances often strengthen rapidly as they move over the zone of rapidly changing temperature along the East Coast. Nor'easters often form in this way along the Mid-Atlantic region, then strengthen rapidly as they move northeast over the coastal waters. Winds blowing counterclockwise around the center of the low pressure area cause strong northeast winds ahead of the advancing storm.

Particularly damaging nor'easters have occurred in March 1962; February 1978; and November 1992. In such severe nor'easters, northeast winds blowing at gale force or higher build huge waves and drive ocean water towards the coast. Tide levels rise many feet above normal level, resulting in flooding of low-lying coastal regions. Battering waves strike beachfront houses and piers, causing significant damage. High waves and wind can imperil ships at sea, while inland heavy rain causes flooding of streams and urban areas. If the air temperature is cold enough, nor'easters can result in crippling snowstorms. The combination of extreme weather and the highly populated coastal regions of the Northeastern United States can result in significant damage whenever a nor'easter strikes. *See also: Air; Atmosphere; Extratropical Cyclone; Hurricane; Jet Stream; Low Pressure; Oceans; Rain; Snow; Temperature; Trough; Weather; Wind.*

Numerical Weather Prediction

Numerical weather prediction is the process whereby computers use complex mathematical equations to predict the future state of the atmosphere given data that describes the initial state of the atmosphere. Numerical weather prediction is an essential part of the weather forecasting process. Meteorologists use the output from computers as guidance in creating the weather forecasts that people see on television, hear on the radio, or read in a newspaper.

The Work of Computer Models

The set of equations used in numerical weather prediction approximates the physical laws that govern the motion of air in the atmosphere. These equations calculate how air moves, and how its temperature and humidity change over time. Other equations used in numerical weather prediction simulate the development of clouds and precipitation, or the interaction between the ocean and the atmosphere. Together, all these equations form a computer model. Models predict the behavior of the atmosphere for hours or days into the future.

Because computers are not powerful enough to compute the motion of every air molecule in the atmosphere, they only predict weather conditions at specific points, called grid points. The grid points in a model are situated evenly across the model's domain. The domain of the model is the part of the Earth for which the model computes weather conditions. A model's domain might be as large as a continent or as small as a state.

One of the most popular computer models currently used by meteorologists predicts weather conditions for grid points spaced 32 km (around 20 miles) apart, and for forty-five vertical layers of the atmosphere, over North America. The spacing of grid points is called the model's resolution. The higher the resolution, the more detail in the weather an atmospheric model can predict, and the longer the model takes to run and process its output. A typical atmospheric model requires two to three hours to compute weather conditions at thousands of grid points at six-hour intervals throughout two days in the future.

The basic process of numerical weather prediction involves computing the model's equations over and over again, each time using the output from the last run-through as input for the next. The process starts with a set of data that describes the state of the atmosphere at a beginning point in time. This set of data is called the initial conditions, and consists mainly of the observed weather conditions over the model's domain. These initial conditions are fed into the model equations, which compute how these conditions will change over a small period of time. The output is then fed back into the same set of equations, so that the state of the atmosphere is predicted for another small time step into the future. This process is repeated over and over again until the computer has predicted weather conditions for hours or days in advance. The final output is then converted from raw numerical data to a form that can be used by meteorologists, such as maps showing the predicted weather patterns. Meteorologists use this computer-generated data as part of their daily forecast decision-making process.

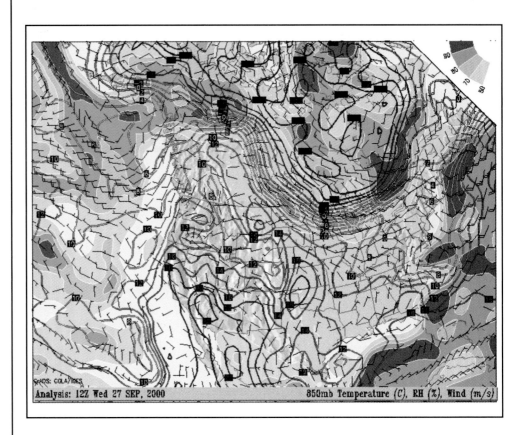

This map shows output from a computer model used by professional meteorologists.

LIMITATIONS OF NUMERICAL WEATHER PREDICTION

There are some important limitations to numerical weather prediction. One of these concerns the set of initial conditions that describes the beginning state of the atmosphere in a computer model. A scientist by the name of Edward Lorenz, working at the Massachusetts Institute of Technology in the early 1960s, discovered that extremely small variations between sets of initial conditions would lead to unexpectedly large differences in forecast output. Numerical weather prediction, therefore, is extremely sensitive to initial conditions. These initial conditions come from actual weather observations taken at thousands of locations around the world. Even though there are many observation sites, they are often spaced far apart. Some parts of the world, such as oceans or deserts, contain few or no observing sites. Initial weather conditions, therefore, are limited in their detail. The computer models do not know what weather is occurring between the initial condition data points, and this lack of data in the initial conditions can result in significant errors in the computer model forecasts.

Another limitation to numerical weather prediction is associated with the equations the computer models use to simulate the atmosphere. While these equations are extremely complex and do a good job of predicting the behavior of the atmosphere, they make approximations and assumptions about various atmospheric processes. These assumptions are made to prevent the model from becoming so complex that it takes too long to compute details in the weather that can usually be ignored. Sometimes, however, these small-scale details contribute to important changes in the atmosphere. Mathematics has not yet developed to a point where equations can fully describe the chaotic behavior of the air.

Finally, numerical weather prediction must not only simulate the atmosphere, it must simulate the land and the oceans over which the atmosphere lies. Weather systems are highly influenced by ocean currents, mountain ranges, and the general distribution of land and water across the earth. Computer models do not exactly simulate the flow of air over the earth because they do not have enough power to exactly represent the earth's complex terrain. Despite the inherent flaws in numerical weather prediction, atmospheric models have become increasingly sophisticated over the past few decades, and the result has been a measurable increase in the accuracy of weather forecasts. *See also: Air; Atmosphere; Oceans; Precipitation; Weather; Weather Forecasting; Temperature.*

OCCLUDED FRONT

An occluded front is a boundary in the atmosphere that marks the division between two relatively cold masses of air. On a weather map, an occluded front is shown by a purple line with alternating triangles and semicircles facing the direction in which the front is moving. Occluded fronts are often accompanied by unsettled weather, including clouds, rain, or snow, though they are generally weak with relatively small variations in temperature and humidity from one side of the front to the other.

Occluded fronts often form when a fast-moving cold front overtakes a slower-moving warm front ahead of it. The cold front will not overtake the warm front all at once but will do so first at a point where the two fronts intersect, usually adjacent to an area of low pressure. As more and more of the cold front collides

An occluded front forms when a cold front overtakes a warm front ahead of it.

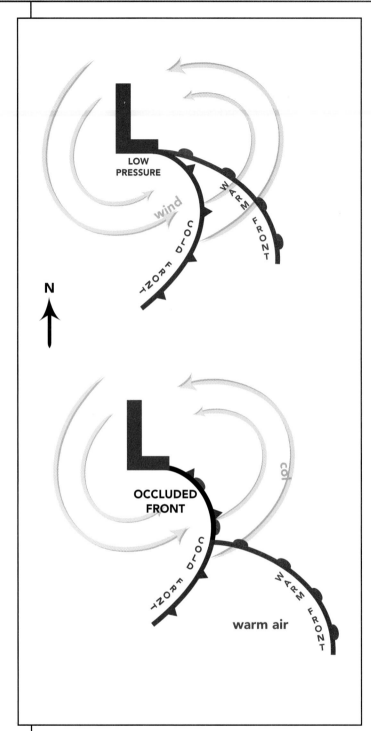

with the warm front ahead of it, this intersection point will move farther and farther south. The occluded front forms north of the occlusion point where the two fronts merge, much like the closing section of a zipper as it is pulled up a jacket. Along the developing occluded front, cold air behind the cold front intersects with cold air ahead of the advancing warm front.

Occluded fronts can either be "warm" or "cold," depending on the relative temperatures of the air behind the cold front and ahead of the warm front. If the air ahead of the warm front is colder than the air behind the cold front, then a warm occluded front will form. If the air behind the cold front is colder than the air ahead of the warm front, then a cold occluded front will form. Air rising along an occluded front carries moisture into the atmosphere, producing clouds and precipitation. Since the temperature difference from one side of an occluded front to the other is not as great as that over a cold front or a warm front, occluded fronts are often difficult to locate on a weather map. ***See also: Air; Atmosphere; Cloud Formation; Cold Front; Low Pressure; Rain; Snow; Temperature; Warm Front; Weather Map.***

Oceans

An ocean is an enormous body of salt water that covers millions of square miles. Oceans cover 70 percent of the earth's surface and hold over 95 percent of all water on the planet. Ocean water circulates around the planet in currents. These currents are often hundreds of miles across and thousands of miles in length, and carry warm or cold water toward the poles or the equator.

How Oceans Affect the Weather

Oceans strongly influence the earth's weather, primarily by providing heat and moisture to the atmosphere. The atmosphere, in turn, also influences the oceans, as winds contribute to the formation of ocean currents. Together, the ocean and the atmosphere are two vast systems of fluid molecular motion that affect one another in complex ways.

Two examples of ocean-atmosphere interaction are El Niño and La Niña, names given to the warming and cooling of ocean water in the eastern Pacific. El Niño and La Niña affect weather patterns around the globe through processes not yet fully understood.

Both air and water are fluids (containing molecules that flow freely in a shapeless substance), and different fluids can affect each other through exchange of heat and momentum. As air blows against ocean water it causes it to move, transferring momentum from air to water. Ocean currents have developed in large part because of the average wind flow around the globe blowing against the ocean surface. These currents spread warm water poleward and cold water equatorward, thus redistributing heat energy around the planet.

Ocean water affects the atmosphere primarily through the exchange of heat and moisture.

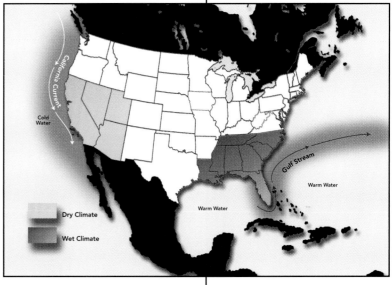

Cold and warm currents influence the climate of nearby land masses.

Warm ocean water heats the air in contact with it, while cold ocean water cools the air in contact with it. Since the pressure of air is a function of temperature, ocean water temperatures can influence weather systems by changing air pressure. A cold ocean current, for example, causes air above it to become colder and denser. The increased air density results in higher air pressure at the earth's surface. It also results in a more stable atmosphere, in which air is less likely to rise. Since clouds and precipitation develop in response to rising air, the climates of land masses adjacent to cold ocean currents are often dry.

By contrast, warm ocean water heats the air above it, making it less dense

163

Oceans cover seventy percent of the earth and influence weather patterns worldwide.

and decreasing surface air pressure. Warm ocean water also evaporates more easily, thereby releasing more moisture into the air. The air above a warm ocean current is therefore warmer, more humid, less dense, and more likely to rise into the atmosphere. The climates of land masses adjacent to warm ocean currents are therefore relatively wet, on average.

EXAMPLES OF OCEAN-ATMOSPHERE INTERACTION

Two specific instances of ocean-atmosphere interaction that are currently the subject of intense study are El Niño and La Niña. El Niño is a warming of ocean water in the tropical eastern Pacific, while La Niña is a cooling of ocean water in the tropical eastern Pacific. These two events alter patterns of wind and pressure across the entire Pacific Ocean. These changes, in turn, influence weather systems in other parts of the world.

Strong El Niño and La Niña events can cause drought in regions that are normally wet and can cause floods in regions that are normally dry. Since the oceans and the atmosphere interact with one another, it is extremely difficult to find original causes of events such as El Niño or La Niña. Scientists investigate how changes in the atmosphere may lead to changes in the ocean, and how changes in the ocean may lead to changes in the atmosphere. ***See also: Atmosphere; Cloud Formation; Drought; Floods; El Niño; Heat Transfer; La Niña; Precipitation; Temperature.***

Orographic Lift

Meteorologists use the term "orographic lift" to describe how air is forced to rise as it blows across sloping terrain or against mountain ranges. Orographic lift is of interest to meteorologists because it can result in clouds, rain, and snow. It can also cause a preexisting area of rain or snow to become heavier. It explains why some mountain ranges are relatively wet on one side and relatively dry on the other.

When air blows up a slope or against a mountain, it is forced to rise simply because the gaseous air cannot blow into the solid earth. The rate of rise may be gradual, as with air blowing up a gradual slope over hundreds of miles. It may be abrupt, on the other hand, if air blows against a steeply rising mountain range. Whatever the process, the rising of air in response to rising terrain is called orographic lift.

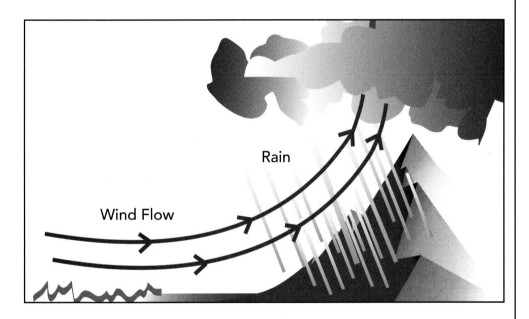

Wind blowing up the side of a mountain range causes rain to fall in the process of orographic lift.

Formation of Clouds, Rain, and Snow

Orographic lift can cause clouds, rain, or snow to develop when the physical properties of air change as it rises. As air gains elevation, it moves into lower atmospheric pressure, since less of the total atmosphere remains to weigh down from above. Moving into lower pressure, the air expands and becomes less dense. The air molecules move around more slowly, causing the air temperature to fall.

Water vapor molecules in the air also slow down. If the air cools enough, large amounts of water vapor molecules will begin to condense. (Condensation occurs when gaseous water vapor molecules transform into liquid water droplets by colliding and adhering to one another.) When condensation occurs in rising air, a cloud forms. The more water vapor there is in the air and the higher the air rises, the more condensation occurs and the bigger the cloud grows. If enough moisture is present in the air, the cloud may eventually produce rain or snow.

Since a steep mountainside can force air upward very quickly over a short distance, orographic lift can produce heavy rain or snow along the side of the mountain range that faces the wind if enough moisture is present in the air. As the air crosses over the mountain and descends the other side, it compresses and warms as it moves into higher atmospheric pressure. Moisture in the air evaporates, and clouds and precipitation dissipate. Mountain ranges, therefore, tend to have much more rain, snow, and cloudiness on the side that faces the prevailing wind flow.

Where Orographic Lift Occurs

An example of orographic lift often occurs along the Sierra Nevada mountain range in California and extreme western Nevada. These mountains are generally aligned from north to south with their western slopes facing the Pacific Ocean. During winter, strong orographic lift can occur as westerly winds blow abundant Pacific moisture against the western-facing slopes of the Sierra Nevada. While lower elevations in California might receive 0.50 to 1.0 inches of rain, locations high in the Sierra Nevada may receive as much as five inches of rain. If the air is cold enough at high elevations, the precipitation will fall in the form of heavy snow. Ski resorts in the Lake Tahoe region in the Sierra Nevada benefit from orographically-produced snowfall. ***See also: Air; Atmosphere; Precipitation; Rain; Snow; Temperature; Water Vapor; Wind.***

Overrunning

Overrunning occurs as relatively warm air flows up and over colder air near the earth's surface. It often develops as southerly winds blow up and over a warm front or stationary front lying from east to west. The result can be a widespread area of clouds, rain, or snow on the opposite side of the front from which the air is blowing.

The term "overrunning" is generally used by meteorologists to describe a warm advection pattern. Warm advection is simply the horizontal movement of relatively warm air through the atmosphere. Since warmer air is less dense than

OVERRUNNING

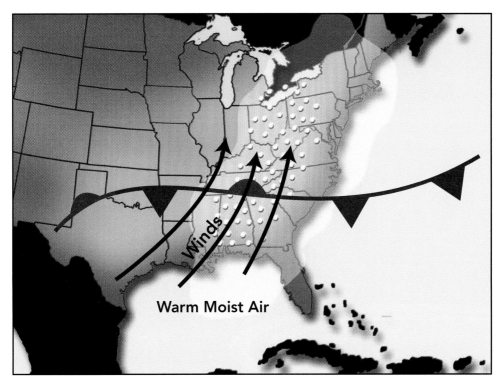

Overrunning occurs as warm moist air blows up and over a front.

colder air, it will rise up and over colder air lying at the surface of the earth. Normally, air rises only gradually as the result of warm advection. However, when warm advection blows air horizontally over much colder air north of a front, the warmer air will rise much more quickly. Rising air changes in density, pressure, and temperature, causing gaseous water vapor in air to transform into liquid cloud droplets. The more moisture in the air and the faster and higher the air rises, the larger the clouds and the more likely the cloud droplets will grow into rain droplets or snowflakes. Sometimes, the southerly winds that cause warm advection and overrunning over a front originate from over an ocean. These winds carry a very high moisture content, so the overrunning results not only in extensive cloudiness along and north of the front but widespread and sometimes heavy rain or snow.

During winter, freezing rain and sleet sometimes develop when overrunning creates an inversion in the atmosphere. Normally, air temperature decreases with increasing height. When an inversion exists, air temperature increases through a layer of the atmosphere. Warmer air flowing up and over colder air near the earth's surface can cause an inversion. For freezing rain or sleet to form, snow first falls into the relatively warm inversion and melts. As it falls back into colder air underneath the inversion near the earth's surface, it refreezes into freezing rain or sleet. In winter, meteorologists know that overrunning patterns in the vicinity of very cold air near ground level may result in dangerously icy conditions. *See also: Advection; Air; Atmosphere; Cloud Formation; Freezing Rain; Oceans; Rain; Sleet; Snow; Stationary Front; Temperature; Warm Front.*

Ozone

Ozone (O_3) is a gas, the molecules of which consist of three oxygen atoms. Ozone is most abundant and plays its most important role in the stratosphere, a layer of the atmosphere extending from around seven miles to thirty miles in altitude. In the stratosphere, ozone absorbs incoming ultraviolet (UV) radiation from the sun, preventing it from reaching the earth's surface. In the mid-1980s, scientists discovered that ozone levels were significantly depleted over the Antarctic. This reduction in stratospheric ozone became known as the ozone hole and continues to be the subject of major scientific research.

Ozone is a naturally occurring gas that exists in very small amounts throughout the atmosphere. Around 90 percent of atmospheric ozone is found in the stratosphere. Even at its greatest concentrations, it is a relatively scant gas, accounting for only up to ten per every million air molecules. Near the surface of the earth, ozone is a pollutant, sometimes causing health problems as it gathers around cities in a stagnant summer weather pattern.

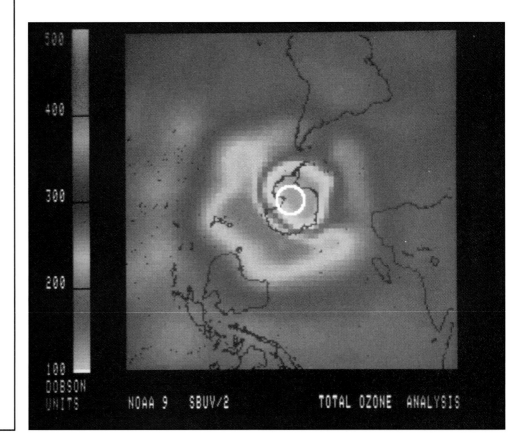

This imagery from a satellite shows the ozone hole over Antarctica.

THE DESTRUCTION OF OZONE

Ozone is created and destroyed through natural processes. It forms when ultraviolet radiation from the sun strikes oxygen (O_2) atoms. This breaks the oxygen molecule into two oxygen atoms; these single oxygen atoms may then recombine with other O_2 molecules to produce ozone. Ozone molecules can break apart when they combine with other ozone molecules, other free oxygen atoms, and nitrogen compounds. Ozone molecules also break apart when they absorb ultraviolet radiation.

In addition to the natural process of ozone destruction, human-produced chemicals called chlorofluorocarbons (CFCs) can also destroy ozone. CFCs were invented in 1928 and eventually became widely used in industry and consumer products such as Styrofoam, propellants in aerosol sprays, and coolants such as those used in refrigerators. Once released into the atmosphere, CFCs drift up to the stratosphere where UV radiation breaks them apart into chlorine (Cl) and other molecules.

Chlorine is a very efficient destroyer of ozone. Scientists have estimated that one chlorine atom can destroy 100,000 ozone molecules over a period of decades. The more CFCs introduced into the stratosphere, the greater the destruction of ozone molecules. Since CFCs last so long, they continue to destroy ozone in the stratosphere long after humans reduce or cease production at the surface.

The destructive power of artificially created chemicals on atmospheric ozone was highlighted in the mid-1980s by the discovery of the ozone hole over the Antarctic. The ozone hole is an area of significantly depleted stratospheric ozone centered over the South Pole and covering the entire Antarctic continent. The ozone hole reaches its maximum each year in the late Southern-Hemispheric winter and early spring, and diminishes through the rest of the year. When the ozone hole is at its peak, 40 to 70 percent of stratospheric ozone can be depleted over the Antarctic.

The size of the ozone hole is restricted by a belt of high winds, called the polar vortex, that forms around the Antarctic each winter and seals the Antarctic atmosphere off from the rest of the planet. As temperatures within this belt over Antarctica plummet to -120°F to -130°F, frozen clouds form in the stratosphere. These clouds not only remove molecules from the atmosphere that would otherwise destroy chlorine but also permit chemical reactions that increase the amount of chlorine. In this way, more and more chlorine builds up through the wintertime darkness over the Antarctic, and ozone is destroyed in greater amounts. When the polar vortex dissipates in the spring and temperatures rise, stratospheric clouds disappear and the high concentration of chlorine diminishes as fresh air mixes in from the north.

Over the Arctic, by contrast, scientists have found no corresponding ozone hole. The main reasons seem to be that temperatures are slightly higher over the Arctic and fewer stratospheric clouds form. Also, the polar vortex is not as strong as it is over the Antarctic, so stratospheric air is able to mix with air from outside the Arctic and prevent chlorine levels from increasing.

CONSEQUENCES OF OZONE DEPLETION

Scientific research has linked long-term exposure to UV radiation with skin cancer, cataracts, immune deficiencies, and damage to microscopic oceanic plant life. Without a sufficient amount of ozone in the stratosphere, life at ground level would be damaged or destroyed by excessive exposure to UV radiation.

Because a reduction in stratospheric ozone poses a significant threat to health, countries all over the world have agreed to ban or limit production of CFCs. The United States banned all nonessential CFCs in the 1970s. An international ban on ozone-depleting chemicals, called the Montreal Protocol, was introduced in 1987. Since then, at least 160 nations have signed the agreement. Meanwhile, scientists continue to conduct research over the Antarctic and elsewhere, using data collected on airplane missions and from orbiting satellites, to further understand how ozone behaves in the stratosphere. *See also: Atmosphere; Radiation; Satellite; Weather.*

POLLUTION

Air pollution consists of unhealthy or damaging human-made and natural molecules in the atmosphere. Pollutants can cause health problems, harm the environment, or simply result in an atmospheric haze, which reduces visibility. Pollution is primarily caused by emissions from industry and vehicles, and as such has been a negative result of the industrial revolution since the late nineteenth century. The following are the main contributors to atmospheric pollution:

CARBON MONOXIDE (CO)

A colorless, odorless gas, carbon monoxide is a result of an incomplete burning of fossil fuels, including gas, oil, and coal. It is mostly associated with automobile emissions and can build up in tunnels and garages. Since it is nearly undetectable by humans, it is particularly dangerous; exposure to CO can rob the brain of oxygen and cause death.

Car exhaust is a major source of air pollution.

Chlorofluorocarbons (CFCs)

A by-product of refrigerants, aerosol propellants, and Styrofoam, CFCs are most destructive in the stratosphere, a layer of the atmosphere extending from around seven miles to thirty miles in the atmosphere. Here, CFCs break down into chlorine, a chemical that is particularly efficient in destroying ozone. Ozone serves a protective role in the stratosphere by absorbing ultraviolet (UV) radiation from the sun and preventing it from reaching the surface of the earth. Exposure to UV radiation has been shown to result in serious health problems.

Nitrogen Oxides (NO_2, NO_3, etc.)

Nitrogen oxides are created through the burning of fossil fuels and industrial emissions. They contribute to smog, which can cause respiratory illness in humans and appear as a dirty, brown haze in the sky. Nitrogen oxides also contribute to acid rain.

Sulfur Dioxide (SO_2)

Also produced from the burning of fossil fuels, including the burning of coal and oil at power plants, sulfur dioxide results in smog and respiratory problems and is also a constituent of acid rain.

Hazardous Air Pollutants (HAPs)

These are dangerous chemicals released into the atmosphere often by accident, as in a chemical spill.

Ozone (O_3)

Ozone is produced when emissions from vehicles and industry react with other chemicals in the atmosphere. These chemical reactions occur much more efficiently in hot weather, so ozone pollution is primarily a summertime phenomenon. It results in haze or smog and respiratory distress, including coughing and shortness of breath.

Particulate Matter

The major cause of haze, particulate matter consists of microscopic particles suspended in the air. These particles can come from smoke, dust, and vehicle or industrial emissions.

Hydrocarbons

Also called volatile organic compounds, or VOCs, these pollutants contain carbon from vapors emitted by gasoline and industrial chemicals. Methane is one of the more significant hydrocarbons. *See also: Acid Rain; Air; Atmosphere; Haze; Ozone; Pollution; Radiation; Visibility; Weather.*

Precipitation

Any form of water or ice, such as rain or snow, that falls from the sky is called precipitation. Other kinds of precipitation besides rain and snow include hail, sleet, freezing rain, and drizzle. Precipitation occurs when the ice crystals and water droplets inside clouds grow large and heavy enough to fall to the earth. Precipitation is part of the hydrological cycle, a process whereby water circulates in liquid, gaseous, and solid form between the earth and the atmosphere.

Precipitation forms inside clouds through two main processes. One is called the ice crystal process, also known as the Bergeron-Findeisen process after the two atmospheric scientists who formulated the process. The other way in which precipitation forms is called the collision/coalescence process.

The ice crystal process describes how precipitation develops from ice crystals inside clouds. It is the primary way in which precipitation is created outside the tropics. To better understand how this process works, it is important to understand how water behaves inside a cloud. Clouds form high in the atmosphere where temperatures are much lower than they are at ground level. Outside of the tropics, temperatures in most clouds are below freezing even during summer. Even so, all but the highest and coldest clouds contain an abundance of liquid water droplets as well as some ice crystals. These liquid droplets are called supercooled because they exist in liquid form below freezing. In fact, studies have shown that, owing to their very small size, supercooled droplets can remain in liquid form to a temperature of -40°F.

In the ice crystal process of precipitation development, tiny ice crystals grow into larger snowflakes or ice particles by attracting water vapor molecules from supercooled water droplets. The water vapor molecules freeze on contact with the ice crystal, causing it to grow larger. Eventually, the ice crystal may break apart into smaller fragments of ice, which then continue to grow at the expense of supercooled water. Or the ice crystal may grow into a snowflake large and heavy enough to fall toward the earth. When the atmosphere is relatively warm, as it is during summer, the snowflakes or ice crystals will melt into rain before they reach the ground. During winter, on the other hand, snowflakes may remain frozen all the way to the surface of the earth. Because of the ice crystal process, therefore, rain usually begins as snow high in the clouds outside of the tropics, even during summer.

Precipitation is the name given to any form of water or ice falling from the sky, such as rain or snow.

Sleet or freezing rain occurs when snowflakes melt as they fall into a layer of relatively warmer air above the earth's surface. If another layer of colder, sub-freezing air exists just below the layer of warmer air, this melted precipitation will refreeze into ice pellets, called sleet, or will freeze on contact with the ground as freezing rain. Hail, on the other hand, forms in thunderstorms when strong air currents suspend large ice particles inside the cloud. As supercooled water droplets collide and freeze onto the ice particles, they grow larger, eventually becoming heavy enough to fall to the earth as a hailstone.

The other way in which precipitation forms is called the collision/coalescence process. It is the primary way in which rain develops in the tropics, and it also contributes to precipitation over middle and high latitudes. In the tropics, the atmosphere is warm enough that the temperature inside most clouds does not fall below freezing. The ice crystal process of precipitation formation is therefore unable to occur. Instead, air currents inside the clouds cause the water droplets to collide with one another. When the water droplets collide, they merge, or coalesce, and grow larger. Eventually, they grow large and heavy enough to fall to the ground as rain. The collision/coalescence process also occurs in the lower parts of clouds in middle and high latitudes where the temperature is above freezing. *See also: Atmosphere; Cloud Formation; Freezing Rain; Hail; Hydrological Cycle; Rain; Sleet; Snow; Temperature; Thunderstorms; Water Vapor.*

Radar

Radar stands for radio detection and ranging. In meteorology, radar is used to detect the location, intensity, and movement of precipitation such as rain or snow. A specialized kind of radar, called Doppler radar, can also sense the direction in which the precipitation is being blown by the wind. Radar can show areas of flood-producing rainfall, for example, or the circulation inside a thunderstorm that may lead to a tornado. It is an indispensable tool for use in warning the public of severe weather.

How Radar Works

Radar is able to detect precipitation by using microwave energy. This is the same kind of energy used in microwave ovens, except that radar uses very low and harmless amounts. Microwave energy is emitted from a radar dish in pulses. The energy spreads away from the radar in a widening arc, sometimes for hundreds of miles. When it encounters an airborne object, such as a raindrop or snowflake, some of the energy bounces back toward the radar. There it is received by the dish in between the pulses of outgoing energy. With each pulse of energy emitted and received, the radar dish rotates slightly on its axis and sends out another pulse of energy. In a short amount of time, the radar sweeps 360° around, gathering data on precipitation from all sides of the radar dish.

The energy received by the radar is analyzed and processed by a computer to determine the location, intensity, and movement of the precipitation. The time it takes for a pulse to bounce back from the rain or snow indicates how far away the precipitation is occurring. The more time it takes for the energy to return, the farther away the precipitation. The strength of the reflected energy indicates the intensity of the precipitation. The greater the number of raindrops or snowflakes, the greater the amount of energy that is reflected back to the radar. Doppler radar also senses the frequency at which the energy returns to the radar dish. Energy that bounces off of precipitation that is moving away from the radar will return with a lower frequency than will energy bouncing off precipitation moving toward the radar. This change in frequency, called the Doppler shift, indicates whether winds are blowing toward or away from the radar.

USING RADAR TO OBSERVE WEATHER

Modern doppler radar is extremely sensitive and versatile. It allows meteorologists to view precipitation in a variety of ways. Computers analyze the intensity of reflected microwave energy and correlate this intensity with precipitation rates. When this data is graphically displayed at a computer terminal, meteorologists can see how quickly rain or snow is accumulating over a given area. This kind of information is particularly valuable in warning the public about potential flooding. Doppler radar can also sense the circulation inside a thunderstorm that may lead to a tornado. It is able to show meteorologists the altitude of the tops of thunderstorms and whether the thunderstorms are likely producing hail or not. It can detect not only rain and snow but also the tiny water droplets in clouds, as well as dust and insects. In fact, when Doppler radar is not being used by meteorologists to

Radars are limited in what they can detect. As microwave energy spreads outward from a radar dish, it moves higher and higher into the atmosphere. The farther out the radar beam strikes an object, the higher in the atmosphere that object is. However, precipitation above or below the radar beam goes undetected.

Information received from radar is also limited to a certain distance. The farther away an object is from the radar, the less detail the radar is able to detect. This is because radar energy spreads apart as it travels outward, much like a flashlight beam widening with distance. The farther away from the radar, the wider the area the radar energy covers. Since the energy is spread out over a wider area, smaller and more distant objects in the atmosphere reflect less and less energy back to the radar dish, and the radar receives less detail of the object.

This radar shows Hurricane Hugo crossing the South Carolina coastline on September 22, 1989.

track precipitation, biologists have used it to follow the migrations of birds. Presently, almost 150 radar sites owned and operated by the government allow meteorologists to track precipitation over 95 percent of the country. *See also: Doppler Radar; Meteorology; Precipitation; Rain; Snow; Thunderstorms; Weather; Wind.*

RADIATION

Type of Radiation	Wavelength
Gamma Rays	Less than 0.0001 μm
X Rays	0.0001 - 0.01 μm
Ultraviolet (UV)	0.01 - 0.4 μm
Visible	0.4 - 0.7 μm
Infrared	0.7 - 100 μm
Microwave	100 μm - 1m
Radio and Television	1m - 1 km

Air is warmer around the equator because solar radiation strikes the earth more directly there.

Radiation is energy in the form of electromagnetic waves, invisible waves of energy that travel at the speed of light. Unlike sound waves or water waves, they do not need a substance, or medium, through which to travel. Radiation is emitted by the movements of electrons. All objects both emit and absorb a certain amount of radiation. Meteorologists study the radiation emitted by the sun, the earth, and the atmosphere. Together, this radiation determines the temperature of the earth and drives the weather systems that move through the atmosphere.

THE PROPERTIES OF RADIATION WAVES

The distance from one crest of a wave to the next is its wavelength. The wavelength of the radiation emitted by an object is determined by the temperature of the object; the hotter the object, the more radiation it emits and the shorter the wavelength. The sun, for example, has an average surface temperature of 10,500°F. The peak amount of radiation energy emitted by the sun has a relatively short wavelength of 0.5 μm, where 1μm (1 micrometer) = 0.000001 meters. The earth, on the other hand, with an average surface temperature of 59°F, emits its peak amount of energy with a longer wavelength of around 10 μm. The human eye is sensitive only to visible radiation, with wavelengths of 0.04-0.7 μm.

Three things can happen to a radiation wave when it strikes another object: it is reflected, scattered, or absorbed. Waves that are reflected bounce back from the object. The amount of radiation that is reflected depends on the size of the object and the surface characteristics of the object, such as color. White objects, for instance, reflect visible light more efficiently than dark objects. A wave is scattered when its path has been altered slightly after hitting an object. Scattering occurs when certain wavelengths of radiation strike very small objects, such as air molecules or dust particles. Finally, when radiation hits an object, it may be absorbed. The amount of absorption depends on the size of the object, its molecular structure, the intensity of the radiation, and the angle at which the radiation strikes the object. Radiation that is absorbed by an object is converted into other forms of energy, including heat.

RADIATION AND TEMPERATURE

The exchange of radiation between the sun, the earth, and the atmosphere regulates the earth's temperature. In general, the temperature of an object will increase if it absorbs more radiation than it emits. The temperature of an object will decrease, on the other hand, if more radiation is emitted than absorbed. In this way, the average temperature of the earth depends on the amount of the sun's radiation that it absorbs, compared to the amount of radiation it emits. The average temperature of the planet that results from this radiation exchange is around 59°F, although much of the planet is warmer or colder. The atmosphere also warms the earth because it emits radiation toward the ground. This process is called the greenhouse effect.

Incoming solar radiation is not absorbed in equal amounts over all parts of the planet, and this imbalance in absorption is responsible for the circulation of air around the globe. In the tropics, the sun's radiation strikes the earth much more directly than at the poles. In this region, the earth absorbs more radiation than it emits, causing the surface temperature to be relatively warm. The opposite is true near the poles, where outgoing radiation exceeds incoming solar radiation. In addition to latitude-dependant temperature differences, ocean and land masses warm and cool at different rates. These differences in temperature result in differences in air pressure. Since air tends to move from high to low pressure, the imbalance in temperature and pressure caused by variations in the amount of absorbed radiation results in the winds that blow across the earth. *See also: Atmosphere; Greenhouse Effect; High Pressure; Low Pressure; Oceans; Scattering; Temperature; Weather.*

RAIN

Rain consists of liquid water drops that fall from a cloud to the ground. It is one of the processes in the hydrological cycle, whereby water circulates between the atmosphere and the surface of the earth. Most rainfall is beneficial, replenishing

A raindrop is typically 1 to 5 mm in diameter and falls at a speed of 6.5 meters (around 20 feet) per second. While raindrops are often depicted as tear shaped, they are actually more rounded in appearance. The bottom of the raindrop becomes flattened as it falls through the atmosphere.

the supply of drinking water and enabling plants to grow. Excessive rainfall, however, especially in a short period of time, can cause destructive floods.

How Rain Forms

Not all clouds can produce rain. Only those of sufficient size contain enough water to produce rainfall. Clouds are produced by rising air currents. As air rises, it cools, causing the invisible water vapor mixed in with the air to condense, or change from a gaseous state to a liquid state. When this happens, tiny cloud droplets form. If enough water vapor is present in the air, and if the vertical air currents are strong enough, the cloud will become large enough to produce rain. The larger the cloud and the more water vapor in the air, the heavier the rain may become.

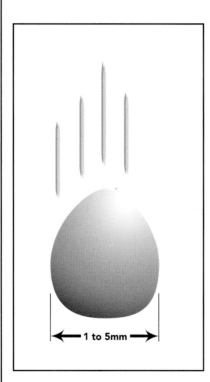

A raindrop deforms into an egg shape as it falls through the air.

Raindrops originate inside clouds through two main processes. In one process, called the ice crystal or Bergeron-Findeisen process, tiny ice crystals inside clouds grow large and heavy enough to fall and melt into rain. The ice crystal process is responsible for most of the rain outside of tropical regions. Even in summer, rainfall begins as ice crystals and snowflakes high inside clouds. The second process that produces rain is called the collision/coalescence process. Tiny cloud droplets grow larger as they collide with one another inside clouds. Once the droplet becomes large and heavy enough it falls to the earth as rain. The collision/coalescence process is the primary way in which rain occurs in the tropics, and it also contributes to rainfall in middle and high latitudes.

Measuring Rainfall

Rainfall is measured by a simple instrument called a rain gauge, which consists of a collecting cylinder set outside away from trees and structures. As rain falls into the collecting cylinder, the amount of rain is measured in inches by marks along the cylinder's side. Light to moderate rainfall can accumulate 0.10 or 0.20 inches in an hour; heavy rain can fall at the rate of an inch per hour or more. Flash flooding occurs when too much rain falls in a very short period of time. Water quickly runs off into streams and creeks that overflow their banks. Flash flooding is usually the result of strong, slow-moving thunderstorms that drop excessive amounts of rain over a small area. ***See also: Air; Atmosphere; Floods; Hydrological Cycle; Thunderstorms; Water Vapor.***

Rainbow

A rainbow is a luminous, brightly colored arc that appears in the sky when sunlight bounces off of raindrops. The different colors in a rainbow are caused by light of varying wavelengths bouncing off of raindrops at different angles. Rainbows can be seen when a person stands with the sunlight at his or her back and looks toward the falling rain.

Rainbows appear from the interaction of sunlight and falling rain.

Rainbows are optical phenomena, resulting from the behavior of sunlight. Sunlight consists of invisible energy called electromagnetic waves, or radiation. Different forms of radiation, including X rays, microwaves, and radio waves, are characterized by their unique wavelengths. The wavelength of radiation is the distance from one crest of the electromagnetic wave to the next. For example, the colors of light that are visible to the human eye each have a different wavelength, ranging from 0.04 to 0.7 μm, where 1 μm = 0.000001 meters. The wavelength of each color of light determines how the light bends or reflects as it encounters different kinds of objects. Sunlight contains the entire spectrum of colors and wavelengths, and together they appear white.

Sunlight splits into the entire spectrum of color after striking raindrops.

When sunlight passes into a falling raindrop, its path bends. This bending is called refraction, and the angle of the refraction depends on the color of the light. Red, which has the longest wavelength, refracts at the smallest angle, while violet, with the shortest wavelength, refracts at the largest angle. When sunlight strikes the drops of water, each of the different colors of light refracts at a slightly different angle, causing the sunlight to split into the entire color spectrum.

A rainbow occurs when the curved interior of the raindrop reflects the spectrum of color back in the direction from which the light came. The angle between the path of light striking the raindrop and the path of light reflected by the raindrop varies from 40 to 42 degrees, depending on the color of the light. When sunlight refracts and reflects back from millions of raindrops, a small spectrum of color emerges from each drop of rain. The raindrops that fall at a 40–42 degree arc from the viewer will reflect this spectrum into the viewer's eye, and it will appear as a rainbow. ***See also: Radiation; Rain.***

Rain Shadow

A rain shadow is an area of relatively low rainfall that lies downwind of mountain ranges. The magnitude of the rain shadow depends of the height of the mountains and their orientation with respect to the average flow of air. Mountains do not

Rain Shadows

create rain shadows by blocking the rain, but rather by their effect on the moisture content of air as it rises and falls over high elevations. Rain shadows are major contributors to the climate of certain parts of the world.

Rain shadows are the result of changes in the moisture content of descending air. As air moves from a high elevation to a low elevation, more and more of the atmosphere weighs down on it from above. The air therefore compresses and the smaller volume causes the air molecules to move faster, warming the air. The water vapor molecules in the descending air also gain speed and will fly apart from one another with greater energy. Any liquid water droplets, such as cloud or rain droplets, will tend to evaporate as the energized water molecules fly apart. As moisture evaporates in descending air, clouds and precipitation dissipate. Rain shadows therefore occur where mountains are aligned perpendicular to moisture-laden winds. Downwind of the mountains, evaporation drastically cuts down on rainfall.

A significant rain shadow is located in the northwestern United States. It lies along the Cascade mountain range, which stretches north-south across western sections of Oregon and Washington. From fall through spring, westerly winds carry rain-bearing weather systems inland from the Pacific. As the abundant Pacific moisture crosses the 8,000- to 12,000-foot peaks of the Cascades and descends the other side, much of it evaporates. While heavy rain and mountain snow occur in western Washington and Oregon and along the west-facing slopes of the Cascades, the eastern parts of these states remain relatively dry. The annual precipitation in lower elevations of western Washington, for example, commonly ranges from 40 to 50 inches. In the Cascade mountain rain shadow, only a hundred miles on the other side of the mountains, the annual precipitation is only 8 to 12 inches. Because of this rain shadow, parts of eastern Oregon are dry enough to be classified as desert. ***See also: Air; Climate; Precipitation; Rain; Snow; Water Vapor; Weather.***

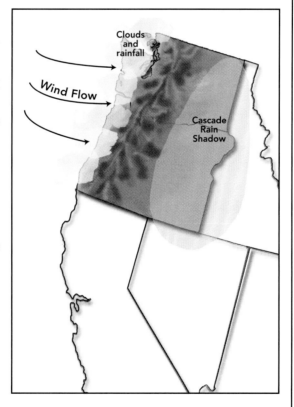

An example of a rain shadow occurs in the Pacific Northwest, east of the Cascade Mountain range.

ROSSBY, CARL-GUSTAV

Carl-Gustav Rossby (1898–1957) was a Swedish-American meteorologist whose work led to major advances in the atmospheric sciences in the twentieth century. Until Rossby's research in the 1930s and beyond, meteorologists tended to study the atmosphere primarily from observations of weather systems at the surface of the earth. Rossby showed not only that surface weather systems coexist with weather systems at upper levels of the atmosphere, but also demonstrated new ways of using physics and mathematics to describe their motions.

Rossby's earliest major contribution to the field of meteorology was his founding of the United States's first university department of meteorology at the Massachusetts Institute of Technology in 1928. Rossby went on to help establish other programs in the atmospheric sciences at the University of Chicago and UCLA in 1940. Rossby also played a major role in the reorganization of the United States Weather Bureau (now the National Weather Service) in the late 1930s. During World War II, he was the leader of the U.S. Army and Air Force's meteorology training program. He also went on to found two professional meteorological journals.

In 1939, Rossby presented a research paper entitled "Planetary Flow Patterns in the Atmosphere," which described a new interpretation of the movement of large-scale weather systems. Rossby's research into the flow of air at high elevations in the atmosphere led to the concept of atmospheric "waves." These waves are defined by northward and southward bends in the flow of air at upper levels of the atmosphere and are created by the migration of large masses of warm and cold air around the planet. In honor of his achievements, meteorologists now call large-scale atmospheric waves "Rossby waves." Rossby also developed the idea that the spin of moving air is an important factor in the development and decay of weather systems. Meteorologists call the measure of spin "vorticity" and use it as an everyday tool in their study of the atmosphere. *See also: Air; Atmosphere; National Weather Service; Meteorology; Weather.*

Satellites

Meteorologists use satellites to take pictures of the atmosphere from high above the earth. These pictures give them valuable information on the location and movement of weather systems. Satellites can sense not only the light reflected from clouds but also invisible atmospheric water vapor and the temperature of both clouds and land surfaces. Using satellite imagery, weather forecasters gain a global perspective on weather patterns. They can see storm clouds over remote oceans or sparsely populated land areas when no other way of detecting these clouds exists. The primary weather satellites now in operation are owned by the United States, Japan, India, Russia, and the European Space Agency.

Types of Weather Satellites

There are two main types of weather satellites: polar orbiters and geostationary satellites. Polar orbiting satellites circle the planet from south to north, or north to south, crossing over or very near the North and South Poles on each orbit.

GOES (Geostationary Operational Environmental Satellite) weather satellite orbits the earth at an altitude of 22,000 miles.

They typically make around fourteen orbits per day at a relatively low elevation of around 700-800 km. As they pass from pole to pole, they take a long strip of images of the cloud patterns below. Since they orbit at a relatively low altitude, the images show a high level of detail. This detail is particularly useful over polar areas, since other kinds of weather satellites are usually positioned too close to the equator to easily view polar regions. A polar orbiting satellite takes hours to complete an orbit and return over a given area to take another photograph. This means that, despite good photographic detail, consecutive images taken by a polar orbiting weather satellite of a given location are spaced farther apart in time.

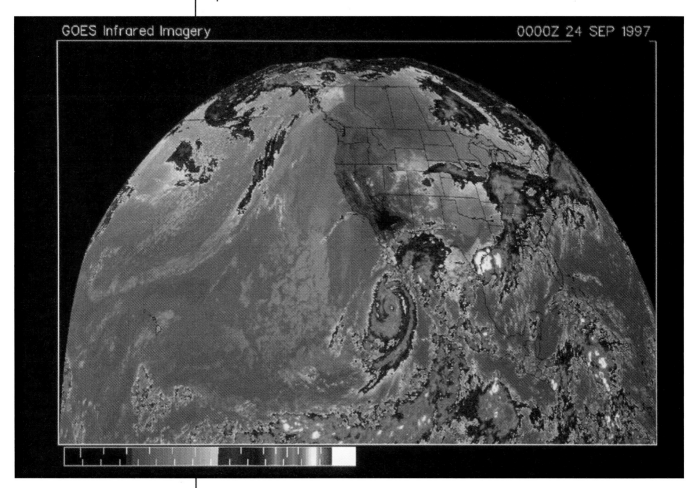

Infared (IR) satilite imagery detects the temperature of clouds. In this IR satellite image, higher, colder clouds show up with darker colors.

Geostationary (GOES) satellites, on the other hand, orbit the earth at a fixed location at low latitudes near the equator. They move in the same direction as the earth, so a GOES satellite is always positioned over the same area. These satellites can therefore take many photographs of the same location every hour, allowing meteorologists to view storms developing rapidly over a matter of minutes. Unlike polar orbiters, GOES satellites orbit at a higher elevation of around 22,000 miles. Since they are situated at such a high elevation, their normal field of view can encompass an entire continent, or even a hemisphere. Thus, a GOES satellite can photograph many locations from a stationary orbit, without the need to be moved. While GOES imagery is not consistently as detailed as that of lower-flying polar orbiters, improvements in technology have led to spectacular detail from recently launched American GOES satellites.

SATELLITE IMAGES

Weather satellites can process three main types of images of the atmosphere. The first is called visible imagery. This kind of satellite imagery is similar to a black-and-white photograph from a camera. Visible satellite imagery shows light reflected off clouds and can therefore be taken only during the day when sunshine is available. It can provide forecasters with high-resolution detail of developing clouds and small-scale atmospheric features.

Infrared (IR) satellite imagery, in contrast, can be used twenty-four hours per day. Weather satellites use sensors to detect the temperature of clouds and of land surfaces. Since the atmosphere becomes much colder with increasing height, clouds show up as very cold objects on IR satellite imagery, while the earth appears relatively warm. Meteorologists use computers to enhance the land-cloud temperature difference, allowing them to see the location and temperature of clouds even during the darkness of night. Since heavy rain and snow often fall from clouds that have high, cold tops, IR imagery is useful in determining the strength of storms.

Lastly, water vapor imagery shows where high and low concentrations of water vapor molecules exist in the atmosphere. Clouds contain high amounts of water vapor and will appear as distinct features on water vapor images. But high amounts of water vapor may also exist even when no clouds are present. Meteorologists look for these areas of atmospheric moisture, as well as areas of dry air, to help them predict the development of storms and forecast the threat of heavy precipitation. *See also: Air; Atmosphere; Precipitation; Rain; Snow; Temperature; Water Vapor; Weather.*

SCALES OF MOTION

Weather systems vary in size and duration, from turbulent eddies that last a few seconds to the vast semipermanent global circulations that pump air from the tropics to the poles. Meteorologists categorize these weather systems using scales of motion. Each scale covers weather phenomena occurring over a given time and area.

The smallest scale of atmospheric motion is the microscale, which covers phenomena that occur over distances of inches to a mile or so. Turbulence, the chaotic flow of air consisting of eddies and whorls that spin and dissolve over a period of seconds or minutes, is a microscale phenomenon. Another kind of microscale phenomenon is called a microclimate. Microclimates are small regions, often linked with outstanding geographical features such as mountain tops, that exist where average weather conditions are substantially different from the surrounding land.

Thunderstorms are mesoscale weather phenomena.

The next largest scale of motion is called the mesoscale. It covers weather systems ranging in size from around a mile to 100 miles or so in diameter, typically lasting from minutes to hours. A thunderstorm is a mesoscale phenomenon, consisting of a cumulonimbus cloud that may cover anywhere from several to tens of square miles and last up to several hours. Another mesoscale phenomenon is lake-effect snow, which often occurs downwind of the Great Lakes. Lake-effect snow consists of small streamers of clouds and snow, often only several miles wide and perhaps fifty miles long, that trail downwind from exposed lake water. They form when cold winds blowing across the lakes draw moisture from the water, which rises into the air and creates clouds and snow.

On the synoptic scale, meteorologists track weather systems that extend for a hundred to a thousand or so miles in diameter and last for hours or days. Cold fronts, warm fronts, and high- and low-pressure areas are all synoptic weather systems. Weather maps often seen on television or in the newspaper, showing positions of fronts and pressure systems, are synoptic scale pictures of the weather across the nation.

The largest scale of atmospheric motion is the planetary scale. It covers weather circulations that exist over thousands of miles, lasting for months or years. The global circulation of air is a planetary weather phenomenon. This circulation transports warm air from the tropics northward, while colder air spreads southward from the poles. In this way, the atmosphere attempts to correct the imbalance in temperature between the cold poles and the warm tropics. ***See also: Air; Atmosphere; Cloud Formation; Cold Front; Lake-Effect Snow; Low Pressure; Microclimate; Snow; Temperature; Thunderstorms; Turbulence; Warm Front; Weather; Weather Maps.***

SCATTERING

When sunlight strikes extremely small objects in the atmosphere, it is deflected in all directions. This deflection is called scattering. Scattering of sunlight is the reason the sky is blue, clouds are often seen as white, and the sun frequently appears red or orange when it sets and rises.

The color of the sky is a result of of the scattering of light off of gas molecules and dust particles.

Scattering occurs when the sun's energy interacts with air molecules, atmospheric dust, and tiny cloud droplets. When radiation from the sun strikes these objects, some or all of the visible radiation is deflected, or scattered, in various directions. Visible forms of radiation have wavelengths (the distance from the crest of one electromagnetic wave to the next) that are relatively small, ranging from 0.40 µm for violet light to 0.70 µm for red light, where 1 µm = 0.000001 meters. Scattering is different from reflection, in that reflected radiation bounces back from an object in the direction from which it came. Scattered radiation, on the other hand, is deflected in all directions.

There are two different kinds of scattering that occur in the atmosphere. The first is called Mie scattering. When sunlight strikes relatively large objects, such as dust particles or cloud droplets, nearly all of the different wavelengths of radiation scatter away from the object. The shorter violet wavelengths scatter along with longer red wavelengths, and all wavelengths in between. When combined together, wavelengths of different colors appear to the human eye as white. Thus, Mie scattering of sunlight from millions of cloud droplets results in a whitish appearance to the clouds. When clouds grow larger, less and less sunlight is able to reach the underside of the cloud, which causes that portion of the cloud to grow darker.

The second kind of scattering is called Rayleigh scattering. It occurs when sunlight interacts with tiny air molecules. Since air molecules are much smaller than cloud droplets, not all wavelengths are scattered. Air molecules scatter bluish wavelengths of light, since these wavelengths are comparable to the size of air molecules, while most other wavelengths pass right through the molecules. These blue wavelengths deflect from the air molecules into a person's eyes, causing the atmosphere to appear blue in color. If no air was present over the earth, no scattering would occur and the sky would appear black, even during midday with the sun directly overhead.

When the sun is low on the horizon, during sunrise or sunset, solar radiation must pass through a much larger section of the atmosphere before reaching a person's eyes. Since the sunlight interacts with a much greater number of air molecules, more and more wavelengths are scattered away. Only the relatively long wavelengths of red and orange are able to pass through such a large section of the atmosphere without being deflected. During these times, therefore, the sun appears as orange or red. Dust in the atmosphere also scatters sunlight and affects the color of the setting or rising sun. On rare occasions, after an exceptionally large volcanic eruption somewhere on the earth, volcanic ash will circle the globe high in the atmosphere. When this happens, the sunlight is not only scattered by air molecules but also by the volcanic ash, often causing sunsets and sunrises to appear unusually red. ***See also: Air; Atmosphere; Radiation.***

Seasons

Seasons are naturally occurring changes in weather and cycles of plant growth that occur as the earth orbits the sun. The driving influence in these changes is the amount of sunlight that reaches the earth, which varies across the Southern and Northern Hemispheres from season to season. As temperatures rise and fall through the seasons, the temperature differences from the tropics to the poles increase or decrease. This has a direct influence on the strength and tracks of low- and high-pressure systems as they move from west to east around the earth.

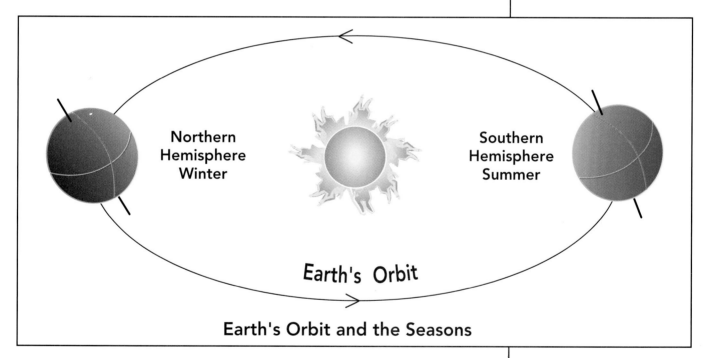

Earth's Orbit and the Seasons

The amount of sunlight that reaches the earth varies as a result of the earth's tilt on its axis. As the earth orbits the sun, its axis is tilted at an angle of 23.5° toward a fixed position in space. During some parts of the year, therefore, the Northern Hemisphere is pointed away from the sun while the Southern Hemisphere is pointed toward the sun. When a hemisphere is pointed away from the sun, it receives less sunlight, and there are fewer hours of daylight. The lower amount of solar energy reaching the hemisphere causes temperatures to fall, sending the hemisphere into the winter season. When a hemisphere is pointed toward the sun, it receives more sunlight and has more hours of daylight.

The season in the northern or southern hemisphere depends on whether the hemisphere is pointed towards or away from the sun.

This corresponds to the summer season, when the air is at its warmest. Therefore, when the Northern Hemisphere is pointed away from the sun during winter, the Southern Hemisphere is pointed toward the sun and is experiencing summer. The opposite is also true: Summer in the Northern Hemisphere coincides with winter in the Southern Hemisphere.

Sunlight shines most directly on the Northern Hemisphere on June 22 and least directly on December 22. However, there is a lag between the time of maximum and minimum sunlight, and maximum and minimum temperatures. The warmest months in the Northern Hemisphere are July and August, while the coldest months are January and February. This is because it takes time for the increasing solar energy in spring to warm the land and ocean. Likewise, the earth takes time to cool in response to dwindling sunlight in fall.

Because of the earth's tilt on its axis, the seasonal change in incoming solar energy varies most in middle and high latitudes, and varies least in the tropics. As a result, changes in temperature from winter to summer are greatest over middle and high latitudes, and least in the tropics. During the winter, the tropics remain relatively warm while middle and northern latitudes grow much colder.

The temperature of the air directly affects its pressure. This is because the higher the temperature, the more quickly the molecules fly around and collide with one another or with surrounding surfaces. These collisions exert pressure, so a higher temperature results in a higher air pressure given a constant density of air molecules. Thus, the increasing difference in temperature between the Northern and Southern Hemisphere in winter corresponds to an increasing difference in pressure. Since wind is the result of pressure differences, the onset of winter strengthens the jet stream, a discontinuous current of wind that blows at high elevations in the atmosphere. In winter, the jet stream also moves farther toward the equator because the temperature difference between the equator and the higher latitudes is greatest during this time. This plays a direct role in the formation and movement of large storms. Winter, therefore, is a time when storm systems become much stronger. In summer, when temperature differences decrease, the jet stream weakens and retreats toward the poles. During this time, storm systems become much weaker and less frequent. *See also: High Pressure; Jet Stream; Low Pressure; Solstice; Temperature; Weather; Wind.*

SLEET

Sleet is a form of precipitation that occurs during colder months. Sleet consists of falling ice pellets, the size of large seeds, which ping and bounce on impact.

Sleet

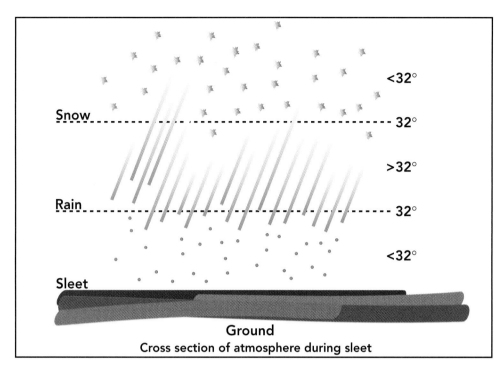

Cross section of atmosphere during sleet

Sleet forms when snow melts into rain as it falls through an inversion, then refreezes before it hits the ground.

It is sometimes confused with hail, a warm weather precipitation that occurs in conjunction with thunderstorms.

Sleet forms when snow falls through a relatively warm layer of the atmosphere. Normally, air temperature decreases with increasing altitude. However, sometimes there exist layers of relatively warm air, with temperatures above freezing, on top of a layer of cold, below-freezing air. Since this is the opposite of how temperature normally varies with height in the atmosphere, the warmer layer of air is called an inversion. If the inversion is thick enough, the snow falling through the inversion partially or completely melts. Studies have shown that at least 300 meters of above-freezing air is required for falling snow to melt into rain.

For sleet to form, the layer of subfreezing air next to the earth's surface must be deep or cold enough for the melted snow to refreeze as it falls. If not, the rain will freeze only after hitting the ground and will coat exposed surfaces in a layer of ice. This is called freezing rain and can be extremely damaging to power lines and trees.

The conditions that support sleet often occur in winter when relatively mild air spreads northward over dense, cold air near the earth's surface. Sometimes local geography aids the formation of sleet, as when mild air overrides cold, dense air trapped in valleys. Sleet often occurs north of an advancing warm front. In these situations, warmer air moving northward rides up and over the cold surface air north of the warm front. Snow melts as it falls through this warmer air north of the front, then refreezes into sleet as it descends into the cold air near the surface. If the warm front is moving very slowly, an extended period of sleet or freezing rain can occur. *See also: Air; Atmosphere; Freezing Rain; Hail; Precipitation; Rain; Snow; Temperature; Thunderstorms; Warm Front; Weather.*

SNOW

Snow is a form of precipitation consisting of ice crystals that fall from a cloud to the ground. A snowflake is a special kind of ice crystal that develops tiny, intricate branches as it grows. Snowflakes can form in clouds when enough moisture is present and the temperature at cloud level is below freezing. For snow to reach the earth, the air temperature must remain near or below freezing as the snowflake falls from the cloud to the ground.

THE FORMATION OF A SNOWFLAKE

Snow originates from tiny ice crystals inside clouds. These crystals can form when cloud temperatures are below freezing. This is a common situation for most clouds outside of tropical regions, even in summer. In order for an ice crystal to form inside a cloud, tiny liquid water droplets must freeze on contact with

Snowflakes are tiny, delicate, and beautiful six-sided crystalline structures.

ice nuclei. Ice nuclei are tiny particles with molecular structures that attract water. Tiny mineral particles suspended in the air are effective ice nuclei.

Once an ice crystal has formed on a nucleus, it attracts gaseous water vapor from nearby cloud droplets. As the water vapor makes contact with the ice crystal, it freezes, and the ice crystal grows in size. Ice crystals therefore grow into snowflakes at the expense of liquid cloud droplets. Because of their molecular structure, ice crystals grow in hexagonal, or six-sided, patterns. The pattern varies depending on the temperature and moisture conditions in which the ice crystal grows. As a developing snowflake is blown around inside of a cloud, up and down through colder and warmer regions, it develops its unique crystalline pattern.

The structure of a snowflake can also change when snowflakes collide with one another. If the temperature is close to, or slightly above freezing, individual snowflakes will stick together to become much larger snowflakes. If temperatures are lower, the snowflakes may shatter into splinters of ice, which may then grow back into more snowflakes.

Degrees of Snowfall

Meteorologists use descriptive terms in weather forecasts to communicate the intensity and duration of snowfall. A flurry is the lightest snow, often lasting for a period of only minutes. A snow shower is a period of usually light snow lasting for less than an hour. A snow squall is an especially heavy snow shower.

Light snow is a steady snowfall that brings little or no accumulation. Visibility in light snow will be greater than 1/2 mile. Moderate snow can bring light to moderate accumulation of an inch or so per hour with visibility greater than 1/4 mile and less than or equal to 1/2 mile. Heavy snow can result in deep accumulation of over an inch per hour, with visibility of 1/4 mile or less. A blizzard occurs when strong winds pick up snow from the ground and whirl it through the air, causing visibility to lower to near zero for hours at a time. *See also: Air; Precipitation; Temperature; Visibility; Water Vapor.*

SOLSTICE

The solstice occurs twice a year, marking the points on the earth's orbit around the sun in which a hemisphere of the earth receives the most and the fewest hours of daylight. One solstice is in winter and the other is in summer. In the Northern Hemisphere, the winter solstice occurs on December 22 and marks the first astronomical day of winter, while the summer solstice occurs on June 22 and marks the first astronomical day of summer. On the winter solstice, the sun's rays are at their weakest and the length of daylight at its shortest. On the summer solstice, the sun's rays are at their strongest and the length of daylight at its

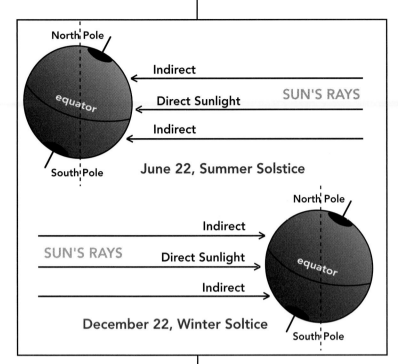

Solstices occur when the direct rays of the sun reach their northernmost and southernmost points.

longest. The solstices mark the changing of the seasons in a regular, predictable way. The weather, however, often does not correspond to the astronomical start or end of a season. Spells of warm weather can occur well before the official start of summer, while cold snaps can occur well before winter begins.

The angle at which the sun's rays strike a given point on the earth changes as the earth rotates around the sun. The earth is tilted on its axis at an angle of 23.5° toward a fixed position in space. The Northern Hemisphere experiences winter when it points away from the sun, while at the same time the Southern Hemisphere experiences summer as it points toward the sun. The situation is reversed during summer in the Northern Hemisphere. The latitude at which the sun's rays strike the earth directly at a 90° angle will change from south to north and back again from season to season. The day on which the sun's direct rays reach their northernmost point, at 23.5° north latitude, is called the summer solstice in the Northern Hemisphere. The day on which the direct light of the sun reaches its southernmost point, at 23.5° south latitude, is called the winter solstice in the Northern Hemisphere. ***See also: Seasons.***

SOUNDING

A sounding is an observation, or set of observations, of weather conditions taken vertically through the atmosphere. Across North America, approximately 100 soundings are taken of the atmosphere twice each day. The information that they provide on the structure of weather systems high above the earth's surface is very important to meteorologists because weather conditions at ground level are strongly linked to weather conditions aloft. The large-scale weather systems, such as fronts and high-pressure and low-pressure areas, extend from the surface of the earth high into the atmosphere, sometimes above 50,000 feet.

THE RADIOSONDE

To get information from high in the atmosphere, meteorologists launch instrument packages from the ground. These instruments are contained in a box called a radiosonde, which is connected to a small gas-filled balloon. When the balloon is launched, it carries the radiosonde high into the sky. As the instruments in the radiosonde ascend through the atmosphere, they take readings of temperature, pressure, wind, and moisture conditions. This data is then sent from the radiosonde to a ground receiving station. Radiosondes can probe the atmosphere as high as nineteen miles before the balloon bursts. The instrument package then parachutes to the ground, often becoming lost or destroyed. Radiosondes are relatively expensive since so much of the equipment is used only once.

USING SOUNDING DATA

Meteorologists can analyze the data from a radiosonde in a couple of ways. One way is to plot the weather conditions on special charts called skew-T diagrams. These charts allow meteorologists to visualize a cross section of the atmosphere by plotting temperature, moisture, and wind readings vertically on lines designating different levels in the atmosphere. The collection of data points on a skew-T diagram represents the sounding made by the radiosonde. Skew-T diagrams are an especially good way for a weather forecaster to determine the stability of the atmosphere over a given area. A stable atmosphere is one in which air does not easily rise, while in an unstable atmosphere, air can rise to great heights. The atmosphere can become unstable if the air temperature decreases rapidly with increasing height. By examining a skew-T diagram, a meteorologist can measure this temperature change and determine just how unstable

A radiosonde is contained in a box attached to a balloon, which is launched into the sky.

195

the atmosphere is. If the atmosphere is unstable enough, large amounts of air can rise and form clouds that eventually will produce rain or snow.

Weather forecasters also examine sounding data by plotting it over maps that show weather conditions on separate horizontal levels in the atmosphere. This allows them to see the flow of air across a large area, as well as locate particular regions of moisture and high and low pressure. Together with observations of weather conditions at ground level, soundings allow meteorologists to obtain a three-dimensional view of the atmosphere.

Sounding data may also be entered into complex computer programs called models that simulate the atmosphere. Twice a day, supercomputers perform complex mathematical equations based on the set of initial observations from a sounding to predict the motions of weather systems. The accuracy of the computerized weather predictions generated from these models is largely a function of the quality of the sounding data fed into it. *See also: Air; Atmosphere; Cloud Formation; High Pressure; Low Pressure; Rain; Snow; Stability; Temperature; Weather; Wind.*

STABILITY

The stability of the atmosphere is a measure of the ease with which air can rise. A stable atmosphere is one that prevents air from rising, while an unstable atmosphere allows air to rise freely. Stability is an important factor in the formation of clouds. An unstable air mass favors cloud formation, which can lead to rain or snow.

When considering the atmosphere's stability, meteorologists often refer to the concept of a "parcel" of air. An air parcel is an imaginary bubble of air that can rise or fall through the atmosphere. As a parcel moves up or down, it does not exchange air molecules with the surrounding atmosphere. It rises or falls as the result of the difference between the density of molecules in the air parcel and the surrounding atmosphere. If the air parcel is less dense than the surrounding atmosphere, it rises, much like a bubble rising through water. If the air parcel is more dense than the surrounding atmosphere, it will sink. The air parcel will become stationary when it reaches a level where the density of the atmosphere is the same as that of the parcel.

The difference in density between an air parcel and the surrounding atmosphere can be determined by the difference in temperature. This is because temperature and density are closely correlated. If air pressure remains constant, then rising air temperature causes air molecules to move faster and spread apart. The density of the air molecules therefore decreases. Likewise, as the air temperature cools under constant air pressure, air molecules slow down and converge, resulting in higher air density.

When the temperature of the atmosphere decreases with increasing height, a rising parcel of air will cool. If the air parcel cools at a faster rate than the surrounding atmosphere, it becomes denser than the surrounding atmosphere and therefore sinks back to a level where the densities are equal. On the other hand, if the rising parcel of air cools at a slower rate than the surrounding atmosphere, it remains less dense than the surrounding atmosphere and continues to rise.

An unstable atmosphere, therefore, is one where the temperature of the atmosphere decreases rapidly with increasing height. Air parcels, once set in motion, continue to rise freely through the atmosphere. A stable atmosphere, by contrast, is one where the temperature of the atmosphere decreases slowly, or even increases, with increasing height. Nudged upward, an air parcel will quickly become denser than the surrounding atmosphere and fall back to its original level.

The stability of the atmosphere is a function of the change in temperature with height of the atmosphere, rather than the rate at which the air parcel cools. The rate at which a rising air parcel cools is constant, determined in part by the behavior of water vapor molecules inside the air parcel. No matter where on the earth or in what situation an air parcel rises, it will cool at the same rate. However, the change in the atmosphere's temperature with height, called the environmental lapse rate, is different from region to region and from day to day. The lapse rate can change if the air warms or cools at low and high levels of the atmosphere. For example, if low levels of the atmosphere warm while high levels of the atmosphere over the same region cool, the lapse rate will increase. An air parcel rising into such an atmosphere would continue to rise because it would remain relatively warmer and less dense than the surrounding atmosphere. If,

Clouds form when air is able to rise.

on the other hand, low levels of the atmosphere become colder while warm air spreads in at high levels, the lapse rate will decrease. This would prevent air from rising, since rising air would quickly become colder and denser than the surrounding atmosphere.

Clouds form when air is able to rise. When the air rises high enough and cools to a certain level, invisible water vapor in the rising air condenses into visible liquid water droplets. These tiny liquid water droplets form a cloud. The more unstable the atmosphere, the higher and faster the air will rise and the bigger and taller the cloud will grow. If the rising air contains enough moisture, the cloud may grow large enough to produce rain or snow. *See also: Air; Air Mass; Atmosphere; Cloud Formation; Rain; Snow; Temperature; Water Vapor.*

Stationary Front

A stationary front is an atmospheric boundary, moving very little or not at all, that separates warmer air from colder air. Typically, colder air will be north of the stationary front while warmer air will be south of the front. On a weather map, a stationary front appears as a series of blue segments with triangles pointing away from the cold air, alternating with red segments carrying semicircles pointing away from the warm air. Stationary fronts may be accompanied by a variety of weather conditions, depending on wind patterns and moisture levels in the vicinity of the front.

Where Stationary Fronts Form

Fronts form along the leading edges of air masses. An air mass is a very large pool of air, often covering thousands of miles, characterized by a consistency of temperature and moisture levels. A stationary front marks the boundary between two air masses that have little or no motion. Like all fronts, rather than being a sharply defined line, a stationary front is better described as a zone of transition. The zone may extend horizontally for several or many dozens of miles, and extend vertically into the atmosphere for thousands of feet. Winds typically blow from the north or northeast on the cold side of a stationary front, and blow from the south or southwest on the warm side of a stationary front. Along the front, air pressure is at a relative minimum.

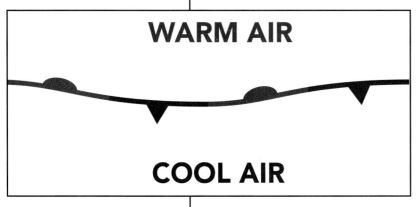

A stationary front marks the boundary between slowly-moving or stalled air masses of warm and cool air.

Effects of Stationary Fronts

While warm and cold fronts can produce characteristic and predictable weather, depending on season and other factors, the weather associated with stationary fronts is much more variable and difficult to characterize. It depends on wind patterns and moisture levels around the front.

If, for example, air currents on either side of the front blow in such a direction as to collide along the front, then air along the front is forced upward into the atmosphere. When air collides in this way, it is called convergence by meteorologists. Convergence can lead to clouds and precipitation along a stationary front if enough moisture is present in the atmosphere. On the other hand, if air currents blow nearly parallel along a stationary front, then convergence will be weak or nonexistent. Air will not rise, and clouds and precipitation will not develop.

Other factors influence weather conditions along a stationary front, such as the presence of weather systems in the upper levels of the atmosphere. The net result is that stationary fronts can be accompanied by clouds, rain and snow, or clear skies.

PROGRESSION OF STATIONARY FRONTS

Stationary fronts do not stay stationary indefinitely. They may dissipate as the contrast in temperature and moisture from one side of the front to the other equalizes. Or they may begin moving again, turning into a cold front if the cold air starts to advance or a warm front if the warm air starts to advance. *See also: Air; Cloud Formation; Cold Front; Convergence; Precipitation; Rain; Snow; Temperature; Warm Front; Weather Maps.*

STORM CHASING

Storm chasing is an activity in which people use vehicles to track and intercept severe thunderstorms and tornadoes. There are two primary purposes of storm chasing. The first is to gather data used in the scientific research of severe storms. In order to better understand these storms, meteorologists must probe the storms with weather instruments from close range. The second is to provide visual observations that confirm indications of severe storms from remote radar stations. This enables National Weather Service meteorologists to issue accurate and timely warnings to the public. Storm chasing can be very dangerous and should be done only by scientists or others who have been thoroughly trained by professionals in the structure and behavior of severe storms.

Storm chasing is primarily an American activity, though some storm chasing is done in southern Canada, England, and Australia. By necessity, most storm chasing occurs in regions where severe thunderstorms are most frequent. Severe thunderstorms are those that produce large hail, damaging wind, and sometimes tornadoes. The central United States, owing to a unique combination of geography and climate, has more tornadoes and severe thunderstorms than anywhere in the world.

Atmospheric scientists chase severe storms to gather data on the structure of the storm, the flow of air through the storm, and the weather conditions that precede storm development. Using this data, meteorologists can simulate the development and behavior of severe storms on powerful computers. Scientists have also developed portable Doppler radars that can be carried in or on a vehicle and deployed in the immediate vicinity of a severe thunderstorm or tornado. The Doppler radar is a powerful tool that enables meteorologists to see in great detail the internal structure of a severe thunderstorm or tornado. The data that these scientists gather in the field is used to better understand how severe storms develop and how they work. A better understanding of the nature of

Storm chasers track down severe weather and relay reports back to meteorologists in weather forecast offices.

severe storms will result in improvements in the ability of weather forecasters to predict and warn for these potentially deadly atmospheric phenomena.

Trained, volunteer citizens, called storm spotters, also chase severe storms. Their purpose is to locate a storm and gain first-hand visual knowledge of the storm's structure and behavior. This is of primary importance for the weather forecasters responsible for the issuance of severe storm warnings. These meteorologists monitor weather radar images that help them determine whether a severe storm is threatening a part of their forecast region. They issue warnings that alert the public to the imminent threat of a severe thunderstorm or tornado. While these warnings may be issued on the basis of radar observations alone, it is only through the direct visual confirmation of a severe storm obtained by a trained storm spotter that the meteorologist is able to know for sure if a threatening storm on radar is actually producing severe weather.

Other kinds of storm chasers include television station personnel looking for dramatic storm video for news shows, and members of the public who, as a hobby, have a great interest in severe storms. Storm chasing can be extremely dangerous, however, and should be done only by meteorologists or people fully trained by meteorologists. Storm chasers frequently drive for many hundreds of miles and many hours, often through very rural parts of the country, with no guarantee of even seeing a storm.

Storm chasers are subjected to the normal hazards of driving, such as traffic and wet, slippery roads. The storm itself poses a variety of hazards. Perhaps the most dangerous is lightning, which severe storms produce in abundance and which kills instantly. Torrential rain from thunderstorms can result in flash flooding. Blinding rain and blowing dust in the vicinity of thunderstorms can reduce visibility to zero, making driving nearly impossible and extremely dangerous. High winds near a severe storm can down trees and power lines. Tornadoes pose obvious danger, and most storm chasers try to remain a mile or so distant from them. Persons not familiar with the often surprising behavior of severe storms can find themselves in a life-threatening situation very quickly.

Scientists and storm spotters use their knowledge of these storms to risk their safety in gathering data that will, over coming years, improve the accuracy of forecasts and warnings. *See also: Air; Climate; Doppler Radar; Lightning; Hail; National Weather Service; Radar; Rain; Thunderstorms; Tornadoes; Visibility; Weather; Wind.*

STORM SURGE

"Storm surge" is a term used to describe the rise in ocean water levels that occurs along a coast as a hurricane moves onshore. It is caused when strong onshore-directed wind piles up water along the coastline. Historically, flooding resulting from storm surge has been the biggest cause of fatalities with hurricanes. Though storm surge flooding is a serious threat, better warnings and evacuations have lessened the casualties resulting from storm surge in recent years.

WHAT INFLUENCES STORM SURGE

Storm surge height is measured in feet as the height to which water rises above the normal tide level. This water rise depends primarily on the strength of the wind blowing the water onshore. Since air spirals counter clockwise around a hurricane, winds on the right side of the system, with respect to its direction of movement, will be directed toward the coast. These winds push ocean water onshore, causing it to pile up and flood low-lying coastal areas. Battering waves on top of storm surge add to the destructive force of the water. Beachfront homes can be completely swept away by storm surge and heavy wave action.

Hurricane Wind Storm Speed [mph]	Surge Height [feet]
74–95	4–5
96–110	6–8
111–130	9–12
131–155	13–18
>156	>19

The stronger a hurricane's wind speed, the higher the storm surge will be.

Three other factors also influence storm surge. The first is the depth of the ocean floor near the coast. The shallower the coastal water, the more water piles up in a storm surge. Another influencing factor is the angle at which the hurricane crosses the coast. A head-on impact with the coastline results in the highest storm surge, while an angular impact lessens the surge effect. Finally, storm surge height is affected by the speed at which the hurricane crosses the coast. Studies have shown that a forward speed of around 30 mph maximizes storm surge. Any faster and the storm outpaces the speed at which water is able to pile up. Any slower and water is able to spread outward laterally along the coastline, lessening storm surge height. The effects of storm surge are worsened if the hurricane crosses the coast during high tide.

EFFECTS OF STORM SURGE

Averaged over decades, storm surge has contributed to the greatest number of fatalities from hurricanes. Indications are that this trend is changing as hurricane warnings and coastal evacuations become more effective in saving lives. In recent years, the main cause of death from hurricanes has been associated with inland flooding from torrential rainfall. However, storm surge remains just as potentially deadly and destructive as ever. Coastal residents should always evacuate when hurricane conditions are expected in their area. ***See also: Hurricane; Oceans; Wind.***

Temperature

The temperature of an object is a measure of its average molecular energy. Meteorologists are concerned primarily with air temperature, which is a measure of the average speed of the air molecules. The higher the temperature, the more energetic the molecules; the lower the temperature, the slower they move on average. Temperature is measured with a thermometer, using one of three different temperature scales.

The most common temperature scale used in America by nonscientists is the Fahrenheit scale, named after the German scientist who developed it in the early 1700s. On this scale, temperatures are measured in degrees Fahrenheit, or °F for short. On the Fahrenheit scale, water freezes at 32° and boils at 212° while an average room temperature is around 70°.

The Kelvin temperature scale is named for the British mathematician and physicist William Thomson Kelvin.

The Celsius scale is the primary temperature scale used around the world outside America. Named after the Swedish astronomer who created it in the mid 1700s, the Celsius scale uses the freezing and boiling points of water as reference points around which the scale is calibrated. Thus, the water freezes at 0°C and boils at 100°C, while average room temperature is around 20°C. All temperature observations, taken every hour or so at thousands of weather reporting sites around the world, are given in degrees Celsius.

The third type of temperature scale is the Kelvin scale, invented by a British scientist in the 1800s. The purpose of the Kelvin scale is to provide a temperature scale that does not include negative numbers. This makes temperature values much easier to use in equations, which makes the Kelvin scale much preferred by the scientific community. Degrees Kelvin are equal to degrees Celsius plus 273. Water freezes, therefore, at 273°K and boils at 373°K, while average room temperature is 293°K.

To convert from °F to °C, use the following equation:
$T_c = 5/9 \times (T_f - 32)$

To convert °C to °F, use this equation:
$T_f = (9/5 \times T_c) + 32$

Where T_f equals the temperature in Fahrenheit and T_c equals the temperature in Celsius. *See also: Air; Thermometer.*

To convert from °F to °C, use the following equation: $T_c = 5/9 \times (T_f - 32°)$	To convert from °C to °F, use the following equation: $T_f = (9/5 \times T_c) + 32°$
Example:	Example:
Where $T_f = 72°$	Where $T_c = 22°$
$T_c = 5/9 \times (72° - 32°)$ $T_c = 5/9 \times 40°$ $T_c = 22.22°$	$T_f = (9/5 \times 22°) + 32°$ $T_f = 39.6° + 32$ $T_f = 71.6°$
$T_c \approx 22°$	$T_f \approx 72°$

THERMOMETER

A thermometer is a device used to measure temperature. Thermometers were first perfected in the 1700s by the same scientists who developed the Celsius and Fahrenheit temperature scales.

KINDS OF THERMOMETERS

The two most popular kinds of thermometers are the liquid-in-glass thermometer and the bimetallic thermometer. Thermometers are sensitive instruments and must be situated with care to be able to provide accurate temperature readings.

A liquid-in-glass thermometer is the most common type of thermometer used today. It consists of a hollow glass tube filled with a colored liquid, usually either mercury or a form of alcohol. The bottom of the tube has a rounded shape and is called the bulb. As the air temperature increases, the liquid inside the bulb expands and rises up the tube. When the air temperature decreases, the liquid contracts and falls back down the tube toward the bulb. Some of these thermometers have devices that automatically mark the highest and lowest point the liquid reaches, allowing the user to read the maximum and minimum temperature for the day.

This liquid-in-glass thermometer shows a reading of 67 degrees fahrenheit.

A bimetallic thermometer measures temperature differently and is distinguishable from a liquid-in-glass thermometer by its clocklike face and circular temperature dial. A bimetallic thermometer consists of two strips of different kinds of metal welded together and formed into a coil. Owing to their different molecular properties, the two metals expand and contract at different rates with increasing and decreasing temperature. This causes the coil to grow larger or smaller as the air becomes warmer or colder. The coil is attached to a pointing device that indicates temperature.

PLACING THERMOMETERS

There are three rules for placing thermometers to get the most accurate temperature reading. The first is that the thermometer must be completely shaded. Sunlight striking a thermometer will cause it to read a temperature much higher than the actual air temperature. Secondly, the thermometer must be placed in a well-ventilated area. Poorly ventilated enclosures allow heat to build up, again causing the thermometer to read too warm. Bank thermometers, often displayed in a large digital readout visible to passing motorists, often suffer from being enclosed in a poorly ventilated box. Finally, the thermometer must be placed at least several feet above the ground. As the sun heats the ground, the air in direct contact with the ground can become much hotter than the air two to three feet above. Thermometer readings taken at stadiums by sportscasters are usually unrepresentative of the actual air temperature because they place the thermometer on the stadium surface, often in sunlight, rather than holding it a few feet above the surface in a shaded area. *See also: Air; Temperature.*

THUNDERSTORMS

A thunderstorm is the combination of rain, lightning, and thunder produced by a cumulonimbus cloud. A cumulonimbus is a very large cloud, extending from several hundred to a thousand or so feet off the ground to as high as 50,000 feet in the atmosphere, and covering an area of perhaps fifty square miles.

FORMATION OF THUNDERSTORMS

Thunderstorms develop when warm, moist air is able to rapidly rise through the atmosphere. For a thunderstorm to form, three ingredients are required: an unstable atmosphere, ample moisture in the air at low levels in the atmosphere, and a triggering mechanism to set the air in motion.

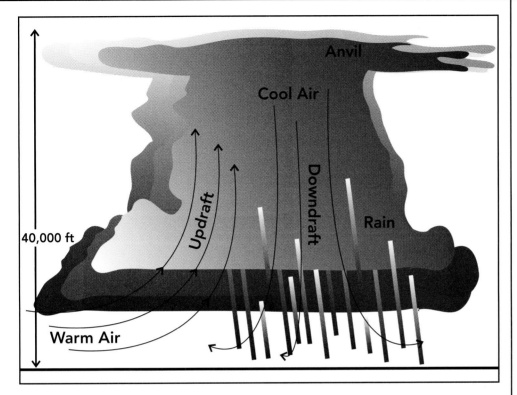

A cross section of a thunderstorm shows regoins of rising and falling air, called the updraft and downdraft.

An example of a thunderstorm trigger is a cold front. A cold front is an atmospheric boundary marking the leading edge of colder, drier, and denser air. This denser air acts like a wedge and plows into warm, moist air ahead of it. This plowing action forces the air to rise all along the cold front, sometimes resulting in a line of thunderstorms hundreds of miles long. Such a line of thunderstorms is called a squall line by meteorologists.

Another example of a triggering mechanism is an area of cold air that spreads in at high altitudes over a region of warm, moist air near the earth's surface. This causes the atmosphere to become very unstable, as the temperature difference between the earth's surface and high altitudes changes rapidly. This type of triggering mechanism often occurs in the Plains states during spring and summer.

Thunderstorms form in an unstable atmosphere. An unstable atmosphere is one in which air is able to rise freely. Thunderstorm clouds are created when a large quantity of warm, moist air rises to great heights. This rising motion is called convection by meteorologists, who use this term synonymously with the word "thunderstorm." Convection is the same process that causes bubbles of heated water to rises through a boiling pot on a stove. The stability of the atmosphere determines how easily and quickly bubbles of air, called parcels by meteorologists, can rise through the atmosphere. Stability can be measured by the difference in temperature between the earth's surface and the air at high altitudes, around 15,000 to 20,000 feet. In general, the greater this difference, the easier it is for warm air to rise.

The second ingredient necessary for the development of a thunderstorm is sufficient moisture in low levels of the atmosphere. As air rises, it cools, causing invisible gaseous water vapor molecules mixed in with the air to change, or condense, into the visible water droplets that form a cloud. As more and more warm, humid air rises through an unstable atmosphere, more and more condensation occurs and the bigger the cloud becomes. Eventually, the cloud becomes large enough and contains enough water to produce rain.

Lastly, there needs to be a process that will set the air in motion vertically. Meteorologists call this process a trigger. The presence of moisture at low levels in the atmosphere in an unstable air mass is often not enough by itself to produce thunderstorms. A trigger is needed to cause convection to occur. A simple example of a trigger is the warming of the ground by the sun. As the ground warms, it heats the air in contact with it, causing the air to become less dense and rise into the atmosphere. As the air reaches its warmest temperature in the afternoon, thunderstorm clouds can grow in scattered locations as large amounts of humid air rise into the sky. This kind of triggering mechanism is common during the summertime across the southern United States.

LIFE CYCLE OF A THUNDERSTORM

Within a thunderstorm, air rises and falls vertically in air currents known as updrafts and downdrafts. The life cycle of an average thunderstorm can be described by the development and decay of these vertical air currents.

Lightning strikes the ground from a thunderstorm cloud in Tucson, Arizona.

In the developing stage of a typical thunderstorm, warm, moist air rises high into the atmosphere from near the surface of the earth in a current called the updraft. A thunderstorm's updraft can be a half a mile to a mile or so wide, and extend vertically for tens of thousands of feet through the thunderstorm cloud. Air cools as it rises in the updraft, causing gaseous water vapor mixed in with the air to condense into tiny water droplets. Many billions of these droplets together create a cloud. The condensation process also adds heat to the thunderstorm, enabling its air currents to rise even higher.

As the updraft carries more and more moisture into the atmosphere, the cloud grows bigger, eventually becoming a cumulonimbus cloud that contains enough water to produce rain. The cumulonimbus cloud becomes electrified as air currents distribute oppositely charged particles to upper and lower regions of the cloud. When this difference in charge builds to a sufficient level, an electrical breakdown occurs and a lightning bolt flashes. At the top of the cloud, the updraft eventually ceases to rise and spreads out horizontally. At this high level, the cloud consists entirely of ice crystals. Winds at upper levels of the atmosphere blow these ice crystals downwind of the updraft, forming the wispy thunderstorm "anvil." The anvil is so named because of its flat appearance.

The mature stage of the thunderstorm begins with the development of the downdraft. The downdraft forms as falling rain drags air down through the cloud to the surface of the earth. Since this air originates from high in the atmosphere, it is much cooler than the warm, moist air flowing into the storm's updraft from the surface of the earth. The downdraft becomes even colder as some of the falling rain evaporates. Evaporation removes heat energy from air, thereby lowering the air temperature and causing the air to become slightly drier. As the falling air becomes colder and slightly drier, it falls even faster, spreading apart horizontally and causing gusty winds as it impacts the surface of the earth. Together, the updraft and the downdraft are called a thunderstorm cell. A thunderstorm may contain one cell or multiple cells in various stages of development. The updraft carries warm, moist air upwards while the downdraft carries cool, moist air downward. In this way, a thunderstorm acts like an atmospheric heat pump.

A thunderstorm dies when its updraft dissipates. This sometimes occurs when the thunderstorm moves into a region of cooler, drier air at the earth's surface. On other occasions, cool air from the downdraft rushes out of the base of the thunderstorm and undermines the updraft, in effect "choking" the updraft off from its source of low-level heat and moisture. Air ceases to rise into the thunderstorm, and the remaining rain inside of the storm falls to the earth. No longer supplied with moisture, the cumulonimbus cloud gradually evaporates as the liquid cloud droplets transform back into gaseous water vapor. Thunderstorms may last anywhere from an hour or two before dissipating, or may last for six hours or more if the thunderstorm updraft maintains its intensity. Often, one of the factors resulting in thunderstorm dissipation is the loss of daytime heating by the sun. This causes low-level air to cool and become denser, thereby making it more difficult for air to rise into the updraft.

Effects of Thunderstorms

Thunderstorms can produce a variety of hazardous weather phenomena, including lightning, flooding rain, high winds, hail, and even tornadoes. Of these, flooding is the leading cause of thunderstorm-related fatalities per year in the United States. Lightning also results in a large number of deaths, especially in parts of the country where frequent thunderstorms coincide with a warm climate that encourages outdoor activities. Tornadoes, while potentially very destructive, are relatively rare.

Thunderstorms produce torrential rain that can lead to flash flooding. Flash

Heavy thunderstorm rains can lead to flash flooding, which can wash away roads.

flooding is so named because of the rapidity with which water rises and overflows rivers and streams. Rain from thunderstorms often causes minor localized flooding of small streams and roads. Sometimes, however, when the atmosphere is particularly humid, thunderstorms produce exceptionally heavy rain. If such a thunderstorm moves very slowly or stalls over an area, locations underneath the storm can receive five or ten inches of rain in a matter of hours. Normally it would require several months to accumulate this amount of rain. In these situations, rivers and streams are overwhelmed and rapidly flow out of their banks. Persons in vehicles are particularly susceptible to the dangers of flash flooding; only a small amount of rapidly moving water is needed to lift a car from the road and sweep it away. Many flash flooding deaths occur when people drive into moving water, not knowing how deep and powerful it is.

Thunderstorms

Lightning occurs after upper and lower parts of a thunderstorm cloud become oppositely charged. When this charge difference becomes too great, a lightning bolt flashes. Many lightning bolts remain inside a thunderstorm cloud. Those that strike the earth, however, can be damaging or deadly. Many lightning fatalities occur when people seek shelter from approaching storms underneath trees. Lightning is naturally attracted to tall objects; when a storm approaches, the best place to be is indoors or in a car. Lightning also causes forest fires that destroy many thousands of acres of trees each year. These forest fires are a natural part of the cycle of forest growth and decay but become particularly destructive when they spread into areas where people have built homes.

Most thunderstorms produce gusty winds that blow outward from underneath the storm. These winds are caused by the thunderstorm's downdraft, a region of air that is dragged down toward the earth's surface by falling rain and hail. Occasionally, a downdraft may become intense, with air rushing down out of the thunderstorm cloud at speeds of 60 mph or more. Intense downdrafts are called downbursts by meteorologists and can snap trees and power lines. The strongest and rarest downbursts have caused winds to gust over 100 mph at the earth's surface. Downbursts also pose a significant threat to aviation. All major airports are equipped with sensors that detect the patterns of wind associated with downbursts and warn incoming aircraft so they can deviate around the storm.

Some thunderstorms produce hail, which consists of falling chunks of ice ranging from the size of seeds to, in rare instances, the size of softballs. Hail can severely damage or destroy crops and accounts for millions of dollars in agricultural damage yearly in the United States. Large hail can dent cars, smash car windshields, and damage roofs of homes and buildings.

The most violent offspring of a thunderstorm is the tornado. Tornadoes typically form underneath the updraft of the thunderstorm, where warm, moist air spirals into a thunderstorm from the surface of the earth. The atmospheric ingredients required for tornado formation do not often come together, and when they are present they require a very delicate balance to be maintained. Tornadoes, therefore, are relatively rare even in the part of the United States where they occur most frequently.

While most hail is small, especially strong thunderstorms can produce large, damaging hail.

THE SUPERCELL

A supercell thunderstorm is a violent, long-lasting thunderstorm that contains a rotating updraft called a mesocyclone. Supercells are severe thunderstorms, meaning they can produce hail three-quarters of an inch or greater in diameter, wind gusts 58 mph or stronger, and/or a tornado. As such they pose a significant threat to life and property. Supercell thunderstorms develop in unstable air when winds are aligned in a certain way from the surface of the earth to high altitudes.

Internally, supercells are distinguished from other thunderstorms by the relation between their updraft and downdrafts. An updraft is an air current directed vertically upward inside a thunderstorm, while a downdraft is an air current directed vertically downward inside a thunderstorm. Updrafts and downdrafts extend for tens of thousands of feet inside a thunderstorm and are often a half mile to a mile or more wide. They distribute moisture and heat vertically inside the thunderstorm cloud. A typical thunderstorm will contain at least one updraft and one downdraft, the pair of which is called a thunderstorm cell.

A supercell contains a rotating updraft called a mesocyclone. Rather than rising straight upward, air spirals up in a corkscrew fashion through a mesocyclone. The updraft gets its spiraling motion from wind shear in the atmosphere. Wind shear occurs when wind direction and speed change with increasing height. Before a thunderstorm develops, wind shear produces circulations of air in low levels of the atmosphere. These circulations rotate about a horizontal axis in the same way a pencil rotates between two moving hands. As air rises in an updraft, it tilts these horizontal circulations into vertical circulations, and the updraft begins to rotate.

A further distinguishing characteristic of supercell thunderstorms is that the updraft is slightly tilted. This is significant because it prevents rain and hail, produced by the rising moisture in the updraft, from falling back into and disrupting the updraft. With a tilted updraft, hail and rain fall out of the thunderstorm cloud both ahead of the updraft and behind the updraft. The tilted updraft acts very much like a fountain of water. When blown by the wind, the stream of water becomes tilted and water falls away from the fountain. When no wind is present, water falls back into the fountain. Winds in the atmosphere must blow in such a way to tilt the updraft over, but must not be so strong that the updraft is tilted so far that it is sheared apart.

The areas of falling rain and hail create two distinct downdrafts in a supercell thunderstorm. Some of the air from these downdrafts spreads out at the earth's surface and interacts with the air rushing into and up through the updraft. This can serve to intensify the rotation inside the updraft. Scientists speculate that this interaction between updrafts and downdrafts in a supercell thunderstorm is an important factor in the formation of tornadoes, which frequently develop in association with a supercell's mesocyclone.

Even though a supercell thunderstorm contains a vast amount of water, dry air is important in its formation for two reasons. Before the storm forms, a layer of dry air at middle levels of the atmosphere, around 5,000 to 7,000 feet in altitude, can act like a cap on the warm, moist air below it. The heat and humidity near the earth's surface will be trapped underneath the layer of dry air, causing more and more heat and humidity to build up over time. Then, when the thunderstorm is finally able to form, it will develop in an explosive manner with an extremely rapid updraft. Air rising through such an updraft can reach speeds of 60 mph or more.

Another way in which dry air contributes to the structure of a supercell thunderstorm is through evaporation. When dry air at middle levels of the atmosphere blows into the backside of the thunderstorm, near the rear downdraft, it causes some of the rain in the downdraft to evaporate. Evaporation causes the air to become slightly drier and also removes heat from the air, causing it to become colder. The colder and drier the air becomes, the denser it gets and the faster it rushes downward through the thunderstorm cloud. As mentioned previously, some of this downdraft air circulates back into the updraft. The speed of air in this rear downdraft must be sufficiently strong to interact with the updraft in such a way as to increase its rotation. It must not be so strong, however, that it cuts off the source of heat and moisture flowing into the updraft. A delicate balance is required. Because of this delicate balance between updraft and downdrafts, a supercell thunderstorm is able to exist for several hours as it moves across the countryside.

FORECASTING THUNDERSTORMS

When forecasting thunderstorms, meteorologists look for the atmospheric ingredients that combine to produce them. These ingredients include warm, humid air at low levels of the atmosphere, an unstable air mass, and a triggering mechanism, such as an approaching weather system, that will initiate thunderstorm formation. Forecasters have a variety of tools at their disposal to analyze the potential for thunderstorm development.

Observations of local weather conditions are very important in determining

Meteorologists use a variety of tools to forecast thunderstorms.

whether thunderstorms are likely to form over a given area. Meteorologists analyze maps of plotted weather observations to see where areas of warm and cold air, atmospheric moisture content, and wind patterns might combine to produce thunderstorms. Surface observations are taken from weather stations around the country at least once per hour. Meteorologists can also look at observations of weather conditions at various levels in the atmosphere above the surface of the earth. These upper-air observations are taken twice per day and sometimes more frequently in regions where thunderstorms are expected to develop.

Upper-air readings are taken by instrument packages called radiosondes that are lofted into the atmosphere on balloons. As the radiosonde rises into the sky, it transmits weather information back to the earth. These upper-air observations are especially useful for forecasting thunderstorms, as they allow forecasters to see how the temperature and moisture of the air varies with increasing height above a given region. The way in which these readings vary can tell the forecaster whether the atmosphere is stable or unstable. An unstable atmosphere is one in which air is able to rise freely. Thunderstorms develop in an unstable atmosphere when warm, moist air quickly rises to great heights.

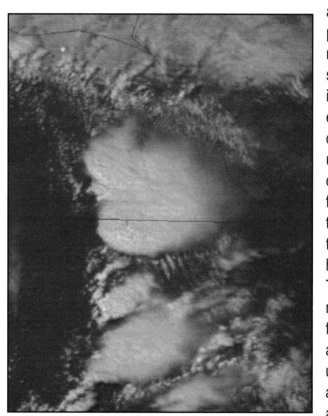

This satellite photograph shows the location of rapidly developing thunderstorms over the central United States.

Satellite photographs enable forecasters to see the movement of weather systems that may act as a triggering mechanism. A triggering mechanism is an atmospheric event that causes air to rise through the atmosphere, building into large thunderstorm clouds. A meteorologist may first consult maps of weather observations to see whether the atmosphere is becoming unstable over a given region. Then, the forecaster may consult satellite photographs to track the movement of an approaching weather system that may trigger thunderstorms to develop in the unstable atmosphere. Forecasters can better estimate the time at which thunderstorms may form by using satellite photographs to judge how fast the weather system is approaching. Satellite imagery can also show exactly where thunderstorms are beginning to form.

Output from computers serves to guide forecasters in their decision-making process. Twice or more per day, large supercomputers run programs called models that simulate the atmosphere and predict weather conditions over large regions for hours or even days in advance. If these atmospheric computer models show thunderstorms forming over a given region, meteorologists can use this information as guidance in creating their own thunderstorm forecast. Meteorologists will use their past experience with the atmospheric model to judge how accurate the model prediction is on any given day.

Radar is an excellent tool for tracking thunderstorms once they have formed. Radar works by transmitting pulses of microwave energy from a large dish. This energy then bounces off of raindrops inside thunderstorms and is received back at the radar dish. This incoming, reflected energy is analyzed by computer to determine the location, strength, and movement of the thunderstorm. Radars can detect thunderstorms as far as two hundred miles from the radar site. They can also detect whether winds inside the thunderstorm cloud are blowing toward or away from the radar site. This information is very useful in determining whether a given thunderstorm is severe or not. Severe storms have pronounced wind patterns detectable by radar.

A special group of meteorologists is devoted to forecasting severe thunderstorms twenty-four hours per day. They are members of the Storm Prediction Center (SPC), located in Norman, Oklahoma, and are a part of the National Weather Service. Using state-of-the-art technology, these forecasters predict and monitor the development of severe thunderstorms and tornadoes across the continental United States. When conditions appear favorable for the development of severe thunderstorms or tornadoes, SPC meteorologists will issue a severe thunderstorm watch or a tornado watch for the threatened region. Typically, a watch will cover an area a couple of hundred miles long by a hundred or so miles wide and will last for six hours. Watches, which are meant to alert the public to the possibility of severe weather, are usually issued an hour before severe weather is expected to develop. Once severe weather does occur, local National Weather Service offices are responsible for issuing severe thunderstorm or tornado warnings. Warnings mean that a tornado or severe thunderstorm is actually occurring and is meant to warn the public to take immediate action to save lives and property. *See also: Air; Air Mass; Atmosphere; Climate; Cloud Formation; Hail; Heat Energy; Lightning; Mesocyclone; National Weather Service; Radar; Rain; Satellite; Stability; Temperature; Tornadoes; Water Vapor; Weather; Wind; Wind Shear.*

Tornadoes

A tornado is a violently rotating column of air extending from the base of a thunderstorm cloud to the surface of the earth. At their most violent, tornadoes contain the strongest winds of any atmospheric phenomenon. Most tornadoes are short-lived and relatively small, though the strongest can last for over an hour and be over a mile wide. Tornadoes strike the central United States more frequently than anywhere else in the world, though a tornado can occur at any location given the right conditions.

Elements of a Tornado

Tornadoes can take on a variety of appearances. The visible portion of a tornado is called the funnel cloud. Sometimes the funnel cloud does not reach the earth, extending only part of the way down from the base of the thunderstorm cloud. However, violent spinning winds may still be causing damage beneath the funnel cloud. In most strong tornadoes, the funnel cloud extends all the way to the earth and can appear in many different shapes. Sometimes it looks like an inverted cone; other times like a cylinder. Most tornadoes are a few hundred yards wide, though the biggest can be well over a mile wide. Others take on a thin, ropelike appearance. While some tornadoes consist of one funnel cloud, others consist of multiple, ropelike funnel clouds writhing around a central tornadic circulation. Meteorologists call these tornadoes multiple vortex tornadoes.

Exactly what atmospheric conditions produce the various tornado shapes is still unknown, though scientists have been able to simulate different kinds of tornadoes in laboratory "tornado chambers." A tornado chamber typically consists of a vented cylindrical or rectangular enclosure several feet high with spinning fans at the top and steam or dry ice at the base. Depending on the rate of movement in the fans and the amount of ventilation, a variety of small funnel clouds can arise from the steam rising from the bottom of the chamber.

When Tornadoes Occur

Most tornadoes are produced in association with thunderstorms. Since thunderstorms are, on average, at their most intense during the warmest part of the day, tornadoes are most frequent during the afternoon and early evening. They have been recorded, however, at all hours of the day, even in the middle of the night.

The wind speed inside a tornado can range from as low as 40 mph to as high as 300 mph. Around 75 percent of tornadoes are relatively weak, with wind speeds less than 110 mph. Less than 2 percent of tornadoes have wind speeds greater than 200 mph. These rare and violent tornadoes, however, are responsible for nearly 70 percent of the tornado fatalities each year in the United States on average. Over 1,000 tornadoes strike the United States each year, causing an average of seventy-five deaths. Tornado-related fatalities have been declining in recent decades because of vastly improved warnings and public awareness.

Because a tornado's circulation is connected to the thunderstorm cloud, the tornado will move along with the thunderstorm. In most cases, this movement is toward the east or northeast at 20-30 mph; less frequently tornadoes approach from the northwest, and even rarer still from other directions. Some tornadoes have been known to race along at 60 mph, while others can be nearly stationary. Tornadoes can remain strong, producing a path of continuous damage as they move along, or they can alternately weaken then strengthen, causing the tornado to appear as though it were skipping. Most tornadoes travel for only a few miles before dissipating, but the strongest can track for many dozens of miles—sometimes over a hundred miles.

Tornadoes are the most violent weather phenomenon on Earth.

Tornadoes often occur as isolated events associated with a single thunderstorm cloud. Other times, however, dozens of violent thunderstorms scattered over hundreds of miles produce numerous tornadoes over a period of hours. These events are called tornado outbreaks and often result in significant damage and casualties. In tornado outbreaks, individual thunderstorm clouds may produce a succession of tornadoes along a track extending for hundreds of miles. One tornado may move along with the thunderstorm for miles before dissipating; then, several minutes later, a new tornado will form underneath the same thunderstorm and continue along for many more miles.

FORMATION OF TORNADOES

Most tornadoes develop in association with a strong, rising current of air, called an updraft, inside a thunderstorm. In particularly intense thunderstorms, called supercells, this rising air current also rotates. Tornadoes usually form inside of or underneath this rotating updraft, known as a mesocyclone. The exact process

A tornado spins underneath the base of a giant supercell thunderstorm over the Texas panhandle.

whereby a mesocyclone intensifies its rate of spin and produces a tornado is not yet fully understood and is the subject of intense scientific research.

The highest frequency of tornadoes on the earth is in the central United States. Here, during spring and early summer, contrasting masses of air and wind can combine to produce severe thunderstorms and tornadoes. One source of air important in the formation of tornado-producing thunderstorms is the Gulf of Mexico. Southerly winds often transport warm, humid air northward from the Gulf into the central United States during the spring and early summer. The moisture in this tropical air mass contributes the enormous amount of water needed to form the massive, tornado-producing thunderstorm clouds.

Cold air is also an important ingredient in the formation of tornadic thunderstorms. Cold air typically spreads in at high altitudes of the atmosphere, anywhere from 17,000 to 30,000 feet, on west to northwest winds. The difference in temperature between the warm air near the earth's surface and the cold air high in the atmosphere causes the atmosphere to become very unstable. An unstable atmosphere is one in which air can quickly and easily rise. Thunderstorm clouds are formed by rapidly rising, moisture-laden air.

In addition to warm, moist air at the surface and cold air aloft, dry air is important in the formation of tornadic thunderstorms. Southwesterly winds blowing from northern Mexico and the desert southwest of the United States carry very dry air into the central Plains. This dry, desertlike air collides and spreads over the humid, tropical air from the Gulf of Mexico. It can suppress thunderstorm development initially, allowing tropical air to build up over hours. Then, like a lid coming off a boiling pot of water, the tropical air can break through the dry air above it, rising into the atmosphere explosively and quickly, forming violent, tornado-producing thunderstorms.

Destructive Tornadoes of the Twentieth Century

This table lists some of the most destructive tornadoes of the twentieth century.

Date	Location	Description
March 18, 1925	Southern parts of Missouri, Illinois and Indiana	Deadliest tornado in U.S. history; raced at 60 mph along a 219-mile track. Killed nearly 700 people; injured more than 2,000.
June 9, 1953	Central Massachusetts	Most devastating tornado in New England history; Worcester, Massachusetts particularly hard hit. Killed 94 people; injured 1,288.
April 11, 1965	Southern Great Lakes,	"Palm Sunday" tornado outbreak; 48 tornadoes strike 5 states. Killed 256 people.
April 3–4, 1974	Midwest Ohio Valley, Southern States, southern Ontario, Canada	"Super Outbreak" of 148 tornadoes in twenty-four hours; biggest tornado outbreak in recorded history. Killed 315 people, injured thousands.
May 31, 1985	Ohio, Pennsylvania, southern Ontario, Canada	Strongest tornadoes ever in this part of the country. Killed 76 people.
May 3, 1999	Central Oklahoma, southern Kansas	Most costly tornado in U.S. history moved through southern Oklahoma City. Damage estimated near $1 billion. Killed 48 people.

Other notable events:

Date	Location	Description
October 3, 1964	Southern Louisiana	Violent tornado spawned by Hurricane Hilda; one of the most damaging hurricane-spawned tornadoes ever. Killed 22 people, injured 165.
July 21, 1987	Teton County, Wyoming	Large tornado tracked for twenty-four miles through the mountainous Teton wilderness, reaching altitudes over 10,000 feet; 15,000 acres of trees flattened. Strongest tornado ever recorded at this altitude.
August 11, 1999	Salt Lake City, Utah	Very rare Utah tornado killed one, injured dozens as it moved through downtown Salt Lake City.

Strong or violent tornadoes, with winds greater than 150 mph, almost always develop in association with supercell thunderstorms. A supercell is an intense, long-lived thunderstorm containing a rotating updraft called a mesocylone. An updraft is a column of rising air, perhaps a half to one mile in diameter, that extends vertically for tens of thousands of feet inside a thunderstorm cloud. It transports warm, humid air from the earth's surface high into the thunderstorm cloud.

A mesocyclone gets its spin from winds in the atmosphere that change speed and direction with increasing height. This change in wind with height is called wind shear. The most widely accepted theory of mesocyclone formation relates the spinning motion in a mesocyclone to wind shear that is present in the atmosphere before a thunderstorm even develops. Wind shear arises when slower winds blowing near the surface of the earth are overlaid by stronger

Tornado funnels can take on a variety of shapes and appearances.

winds blowing higher in the atmosphere. This causes air in low levels of the atmosphere to roll about a horizontal axis, much like a pencil rolling between two moving hands. When a thunderstorm cloud forms, it rises into one of these horizontal rolls. As it rises, it tilts the horizontally rolling air into a vertically oriented column of spinning air. As air rises into this column, it rotates, thus becoming a mesocylone.

A tornado forms when the rate of spin in a mesocyclone increases dramatically near the surface of the earth. How this exactly happens is not yet fully understood. One theory involves the interaction between the mesocylone with nearby regions of descending air called downdrafts. A supercell thunderstorm contains two downdrafts, one ahead and northeast of the mesocyclone and another behind and west of the mesocyclone. Downdrafts form when rain and hail falling out of a thunderstorm cloud drag cool air downward. When a downdraft hits the earth, it spreads out horizontally. Some of this cool air spreading out from the downdrafts circulates back into the warm, moist air rising into the mesocylone. Scientists speculate that the interaction of warm, moist rising air and cool, dense downdraft air can cause the rate of rotation inside a mesocyclone to increase dramatically, causing a tornado to form. The downdrafts must be of a certain strength, however; too strong and the cool downdraft air will disrupt or cut off the flow of warm, moist air into the mesocylone. The balance between inflowing and outflowing air is so delicate that meteorologists now think that fewer than 50 percent of mesocylones in supercell thunderstorms produce tornadoes.

Effects of Tornadoes

A tornado can produce a variety of damage, depending on its intensity. Tornado strength is measured on the Fujita scale, or F-Scale, devised by Theodore Fujita (1920–1998), a professor of meteorology at the University of Chicago. The Fujita scale relates the tornado damage to its wind speed. It ranks tornadoes from zero to five, with five being the most intense.

Tornadic winds cause damage in two ways. One way is that, through sheer force, tornado winds lift roofs off of buildings, collapse walls, roll and toss cars, and uproot or snap trees. The second way tornadic winds can cause damage is by generating large, swirling debris. Tornado debris consists of pieces of buildings, trees, and any other object in the tornado's path, whirling like missiles at over 100 mph. Contrary to popular belief, however, tornadoes do not cause buildings to explode through a rapid loss of air pressure. Instead, buildings break apart because of the force of the winds inside a tornado.

A tornado is ranked on the Fujita scale by the maximum damage intensity found along its path. In many instances, the area of strongest damage may

The Macmillan Encyclopedia of Weather

Violent tornadoes, such as the one that struck Moore, Oklahoma, in April 1999, can cause immense damage.

extend for only a small fraction of the tornado's path, while the rest of the path may contain lesser damage. Therefore, a given tornado ranking does not mean that the tornado produced damage at that intensity for its entire path. Trained meteorologists will survey the damage and assign the F-scale ranking depending on the intensity of the damage they observe. Assessing tornado damage and assigning an F-scale ranking is difficult when the tornado moves through open country. Even a strong or violent tornado can move through open farmland and cause little damage other than to crops.

The following is a listing of the six categories of tornado strength on the Fujita scale and corresponding descriptions of damage.

Fujita Scale - Measuring Tornado Strength

Fujita Scale	Wind Speed [mph]	Tornado Strength	Typical Damage
F0	40–73	Light	Chimneys knocked over, branches torn from trees, windows shattered, billboards blown down.
F1	74–112	Moderate	Mobile homes overturned, outbuildings and barns blown down, some trees and power lines blown over, sections of roofs stripped away.
F2	113–157	Considerable	Roofs blown off, walls damaged. Weak, wood-framed structures, including mobile homes, destroyed. Large trees snapped or uprooted. Small vehicles blown off highways. Some debris missiles generated.
F3	158–206	Severe	Well-built homes heavily damaged, with roofs torn off and exterior walls collapsed. Forests flattened, train cars overturned, vehicles lifted off ground and tumbled. Significant debris missiles.
F4	207–260	Devastating	Well-built homes destroyed, leaving only a pile of debris. Cars hurled and rolled. Forests flattened and trees debarked. Heavy objects turned into missiles. Steel-reinforced or concrete structures heavily damaged.
F5	261–318	Incredible	Well-built two-story homes destroyed and completely swept away leaving only a foundation. Steel-reinforced or concrete buildings damaged or destroyed. Autos fly for hundreds of yards. Asphalt stripped from roads.

The Fujita Scale classifies tornadoes with six levels of wind speed, strength, and damage.

FORECASTING TORNADOES

Meteorologists are able to identify the weather conditions favorable for tornado development and to forecast when these conditions may be present over a given region. However, meteorologists are currently unable to forecast exactly where and when a tornado will strike. Forecasters use satellite photographs, radar, output from computers, and weather observations at the earth's surface and high in the atmosphere to determine if the potential exists for tornadoes over a given area.

The first tornado forecast was made on March 25, 1948, at Tinker Air Force Base near Oklahoma City, Oklahoma, by Air Force Captain Robert C. Miller and Major Ernest J. Fawbush. Five days earlier, a tornado had unexpectedly roared across the base, causing significant damage. As a result of the damaging and unpredicted March 20 tornado, the Air Force directed Miller and Fawbush, the Tinker Air Force Base staff meteorologists, to analyze the weather data from that day to determine if there was an identifiable pattern to the weather conditions that led to the tornado.

Five days later on March 25, they noticed a striking similarity between weather conditions on that day and those from five days earlier, and issued the first-ever tornado warning. Despite infinitesimally small odds, Tinker Air Force Base was indeed struck by another damaging tornado as thunderstorms moved through that evening. The great success of this forecast led to further investigation into the weather patterns favorable for tornado development. Eventually an organization was formed with the sole purpose of forecasting and monitoring severe weather and tornadoes across the United States twenty-four hours per day. Today, this organization is called the Storm Prediction Center, part of the National Weather Service, and is based in Norman, Oklahoma.

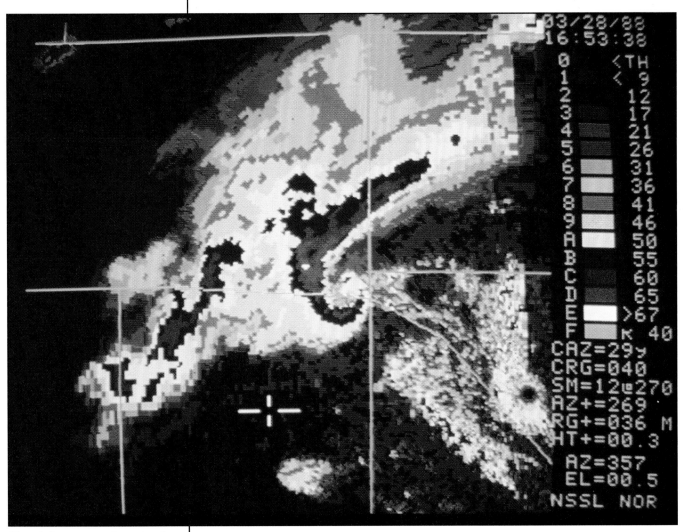

The spiraling pattern at the center of this Doppler radar image is called a hook echo, and marks the location where a tornado is likely occurring.

The weather conditions first identified by Fawbush and Miller are those that meteorologists still look for today in forecasting the potential for tornadoes. They include ample warmth and humidity in low levels of the atmosphere, which serve to fuel thunderstorm development. Cold air is present at high altitudes, increasing the temperature contrast between low and high levels of the atmosphere. The greater this contrast, the more unstable the atmosphere becomes.

Thunderstorms, which produce tornadoes, become much stronger the more unstable the atmosphere is. Often, dry air is present in middle levels of the atmosphere, around 7,000 feet or so. This dry air prevents thunderstorms from forming initially, until a large amount of warmth and moisture has built up near the earth's surface. Then when thunderstorms do develop, they rise explosively into the atmosphere.

Finally, a very important ingredient forecasters look for when predicting tornadic potential is wind shear. Wind shear is the change in wind speed and direction with increasing height through the atmosphere. Wind shear causes air flowing into the thunderstorms to rotate in a corkscrew fashion. Tornadoes usually form inside or underneath this rotating part of a thunderstorm cloud.

When meteorologists at the Storm Prediction Center (SPC) see these conditions coming together over a region, they issue a tornado watch. When a tornado watch is issued, persons inside the watch area should keep an eye to the sky and watch for threatening weather conditions. The watch area is typically rectangular, a few hundred miles long and a hundred or so miles wide. SPC meteorologists try to issue watch boxes an hour before tornadoes are expected to develop. Tornado watches last for up to six hours.

When a tornado has actually been sighted or identified on radar, local National Weather Service meteorologists issue a tornado warning. A warning is much more serious than a watch. Persons in the warned area should take cover immediately, preferably in a basement underneath some sturdy furniture. If a basement is not available, people should take cover in an interior room such as a closet or bathroom on the lowest floor of the building. Mobile homes should always be abandoned when threatened by a tornado since they are easily flipped even by relatively weak tornadoes. People should abandon cars and seek shelter in a building. *See also: Air; Air Mass; Atmosphere; Cloud Formation; Theodore Fujita; Hail; Mesocyclone; Meteorology; National Weather Service; Radar; Rain; Satellite; Stability; Temperature; Thunderstorm; Weather; Weather Forecasting; Wind; Wind Shear.*

TRADE WINDS

Unlike the middle and northern latitudes, where the prevailing wind flow is from the west, winds in tropical regions blow from the east. These tropical easterly winds are called the trade winds, so named because trading ships used them centuries ago to travel from east to west across the oceans. Trade winds are part of the global circulation of air that occurs at lower and upper levels of the atmosphere between the equator and the poles.

The trade winds occur in between two large atmospheric weather systems. One is the region of high-pressure areas, called subtropical highs, that exists on a semipermanent basis around 30°N and 30°S. Air inside these subtropical highs

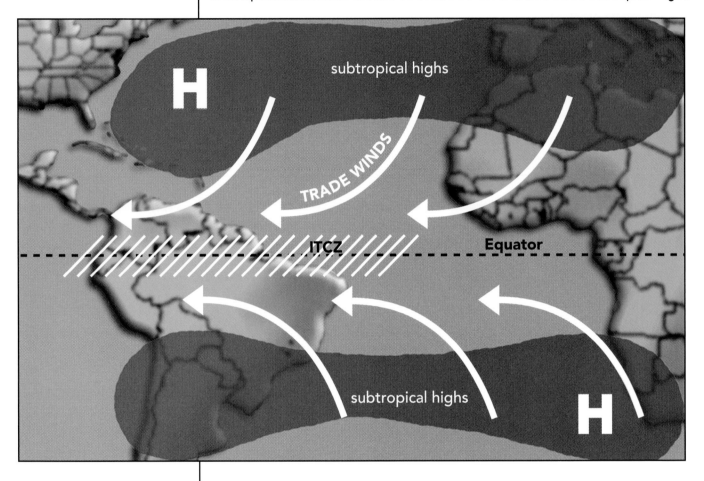

The trade winds blow towards the equator from subtropical high-pressure areas.

sinks from high levels of the atmosphere to the earth's surface. The second weather system is the intertropical convergence zone (ITCZ). The ITCZ is a broad belt of warm, rising air that circles the earth near the equator.

Air sinking in the subtropical highs spreads out horizontally when it hits the earth's surface. Some of this spreading air blows toward the equator and the ITCZ. Because of the rotation of the earth, these winds do not blow directly toward the equator but from an angle. In the Northern Hemisphere, they blow from the northeast toward the ITCZ, while in the Southern Hemisphere, they blow from the southeast.

The trade winds are part of a vast circulation of air that blows for thousands of miles, north and south, at low and high levels of the atmosphere. As air rises in the ITCZ, the trade winds spread in horizontally underneath the rising air to replace it. Fed by the trade winds, the air in the ITCZ rises through the troposphere. The troposphere is the lowest layer in the atmosphere, extending as high as seven miles, in which most weather occurs. Above the troposphere is a layer of stable air called the stratosphere. The stratosphere is referred to as a stable layer of air because it prevents air from rising through it. When air rising through the ITCZ hits the stratosphere, it spreads out horizontally back toward the poles. It eventually descends to the earth's surface in the subtropical highs and returns in the form of the trade winds back toward the equator and the ITCZ. *See also: Air; Atmosphere; High Pressure; Intertropical Convergence Zone; Oceans; Wind.*

TURBULENCE

Turbulence is the chaotic, swirling motion of air over a relatively small distance. It is produced when wind interacts with fixed obstacles, or when wind speed and direction rapidly changes with increasing height in the atmosphere. Turbulence results in the breezes and gusts of wind commonly experienced at ground level, especially on sunny days. Turbulence at high altitudes in the atmosphere can result in a bumpy ride on an aircraft, and in rare instances can pose a threat to aviation.

Most of the turbulence produced near the surface of the earth results when flowing air interacts with fixed objects on the ground. As wind encounters an object, it blows around and over the object's surface. This curving wind flow bends back on itself downwind of the object, forming eddies. An eddy is a small-scale spiraling current of air, similar to the small swirling currents seen in rivers near rocks and other impediments. Eddies form, spin, and dissipate in a chaotic way, meaning that their motion is unpredictable. Eddies forming and dissipating around trees, buildings, hills, and other objects are often the source buffeting wind gusts and breezes at the surface of the earth.

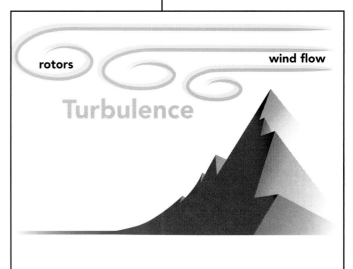

Wind blowing over mountain ranges can cause turbulent eddies called rotors.

Eddies can also form higher in the atmosphere, at altitudes up to 40,000 feet, as air flows over mountain ranges. Like wind blowing around smaller objects at the earth's surface, wind blowing over mountains develops into eddies. These eddies roll about a horizontal axis parallel to and downwind of the mountain range. These horizontal rolls of air are called rotors, and can become very large and strong if the wind speed is high enough and oriented at a favorable angle to the mountains.

Another way turbulence develops aloft is through the interaction between two layers of air blowing at different speeds and directions. The change of wind speed and direction with increasing height is called wind shear. For example, one layer of air at 35,000 feet may be blowing from the west at 50 mph while below it, another layer of air at 30,000 feet may be blowing from the southwest at 30 mph. Turbulence will erupt in between these two layers of air as air molecules collide and swirl in different directions.

Whether created by wind shear or by mountains, turbulence at high elevations in the atmosphere often occurs in clear air. It is therefore referred to as clear air turbulence. When planes fly through clear air turbulence, they often

Turbulance associated with thunderstorms can cause clouds to swirl and mix in chaotic patterns.

experience a bumpy ride. On very rare occasions, the turbulence will be violent enough that a plane will suddenly be thrown up or down thousands of feet, causing injuries to passengers. For this reason, when flying on a plane, it is recommended that passengers keep their seatbelts on—even in clear and smooth air.

See also: Air; Atmosphere; Cloud Formation; Wind; Wind Shear.

Upper-Level Ridge

A ridge is an area of relatively high air pressure. In upper levels of the atmosphere, from 15,000 to 30,000 feet, ridges are defined by a large bend in the flow of air, first toward the pole then back toward the equator. Winds blow strongest around the periphery of the ridge. The ridge itself consists of air at a relatively high pressure and temperature. Like all weather systems in upper levels of the atmosphere, upper-level ridges have a major influence on weather systems at the earth's surface. Typically, though not always, upper-level ridges are associated with clear and tranquil weather at the earth's surface.

Locations underneath upper-level ridges often experience an extended period of fair weather.

Upper-level ridges are easily identifiable on weather maps showing wind speed and direction. The jet stream, a discontinuous current of air blowing from west to east around the planet, bends poleward then back toward the equator around an upper-level ridge. The large poleward bulge in the jet stream defines the northern extent of the ridge. Ridges can extend for hundreds or thousands of miles in length.

Relatively warm air at high altitudes inside the ridge stabilizes the atmosphere. A stable atmosphere is one that prevents air from rising. Since clouds and precipitation develop as a result of rising air, regions over which the atmosphere is stable are relatively cloud-free. Upper-level ridges, therefore, are often accompanied by fair weather.

Ridges generally move from west to east with the prevailing flow of air. However, exceptionally large ridges often become stationary, or even move slowly toward the west against the flow of air. Meteorologists call this backward movement "retrogression." Stationary or retrograding ridges can block the normal progression of weather systems from west to east, forcing them northward or southward around the periphery of the ridge.

When an upper-level ridge stalls or retrogrades, an extended period of stagnant weather conditions can persist below the ridge. These areas may experience little or no rainfall for many days in a row. Other locations, often just west of the ridge, may have repeated periods of rain that can lead to flooding. This happens if southerly winds on the west side of the ridge carry moisture around the ridge northward into a region for a prolonged period of time. As storm systems approach from the west and encounter the moist flow of air on the west side of the upper-level ridge, they can produce copious amounts of rain. As the ridge remains stalled, incoming storm systems repeatedly encounter this moist air, resulting in frequent periods of heavy rain over the same area. ***See also: Atmosphere; High Pressure; Jet Stream; Precipitation; Rain; Weather Maps.***

Upper-Level Trough

A trough is an area of relatively low air pressure. In upper levels of the atmosphere, from 15,000 to 30,000 feet, troughs are defined by a large bend in the flow of air, first toward the equator then back toward the pole. These winds blow along the periphery of the trough, which contains air at a relatively low pressure and temperature. Like all weather systems in upper levels of the atmosphere, upper-level troughs have a major influence on weather systems at the earth's surface. Typically, though not always, upper-level troughs are associated with unsettled and sometimes stormy weather at ground level.

Upper-level troughs are easily identifiable on weather maps showing wind speed and direction. The jet stream, a discontinuous current of air blowing from west to east around the planet, bends equatorward then back toward the pole around an upper-level trough. The large equatorward bend in the jet stream

Upper-Level Trough

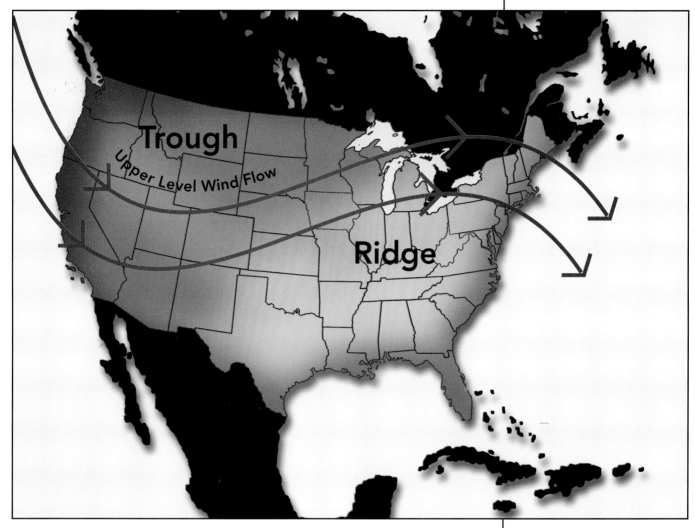

Troughs are located where upper-level winds bend toward the Equator.

defines the southern edge of the trough. Troughs can extend for hundreds or even thousands of miles in length.

Relatively cold air at high altitudes inside the trough contributes to an unstable atmosphere. An unstable atmosphere is one in which air easily rises. Rising air carries moisture to great heights in the atmosphere, causing clouds to form. If the air is unstable enough, and if enough moisture is present in the air, the clouds can eventually grow large enough to produce rain or snow.

Upper-level troughs are often associated with areas of low pressure at the surface of the earth. The surface low-pressure areas typically form just ahead of and underneath the advancing upper-level trough. Meteorologists monitor the development and movement of upper-level troughs on charts of the wind flow and temperature at high altitudes in the atmosphere. Generally, weather conditions will become unsettled ahead of and underneath an upper-level trough as the cold air associated with the trough causes the atmosphere to become unstable. ***See also: Air; Atmosphere; Jet Stream; Low Pressure; Temperature; Rain; Snow; Weather Maps.***

Urban Heat Island

An urban heat island is an area of relatively warm air generated by and centered over a city. Urban heat islands are caused by heat from buildings and vehicles, by the molecular properties of concrete and asphalt, and by a lack of vegetation in urban areas. Urban heat islands are most pronounced at night, when build-

The lack of vegetation in urban areas contributes to the formation of an urban heat island.

ings and roads only slowly release the heat absorbed by the city during the day. Buildings and roads are two of the main contributors to the urban heat island effect. Concrete, asphalt, and stone absorb more heat from the sun than grass and trees. Buildings increase the amount of heat that a city absorbs by increasing the area covered by heat-absorbent materials. Buildings also generate heat from the emissions of cooling and heating systems. Emissions from industries and vehicles also add heat to the atmosphere around cities. Another contributor to higher air temperatures around cities is the lower number of plants and trees. Plants transport water from their roots up through the stem or trunk where it emerges from tiny openings on the undersides of leaves. This moisture then evaporates into the air. The process of evaporation removes heat from the atmosphere, cooling the nearby air. Rural and suburban areas are cooled by the evaporation of water from plants, including grass and trees, as well as evaporation from ponds, lakes, streams, and rivers. In the city, where most of the land is covered with concrete, asphalt, and buildings, this natural cooling process is greatly diminished.

The effects of an urban heat island are much more pronounced at night when buildings and streets release absorbed daytime heat much more slowly than grass, earth, and trees. This slows the nighttime temperature drop over cities compared to suburbs. Because of the differing rates of cooling, nighttime temperature differences between cities and outlying areas can be as much as 20° or more.

Recent studies have shown that urban heat islands can contribute to an increase in rainfall near and downwind of cities. As air becomes warmer over cities, it becomes less dense and rises more easily. Rising air carries moisture into the atmosphere, forming clouds. If enough moisture is present in the air, the clouds can grow large enough to produce rain. This increase in precipitation is most noticeable on warm, summer days when air is already rising through the atmosphere. The added warmth over urban areas gives an extra lift to the air, causing clouds to grow more quickly and to greater heights, thus enhancing rainfall. ***See also: Air; Atmosphere; Cloud Formation; Precipitation; Rain; Temperature.***

VISIBILITY

Visibility is a measure of the maximum distance at which a person can see objects clearly. Rain, snow, fog, and other atmospheric phenomena reduce this distance, measured in miles. In the clearest air, visibility is restricted only by the horizon of the curving earth. The lowest visibility, such as that which occurs in dense fog, is called zero visibility. In zero visibility, objects only yards away cannot be clearly seen. This is an extreme hazard to automobile, plane, and boat traffic.

FACTORS THAT REDUCE VISIBILITY

There are several atmospheric phenomena that reduce visibility. Pollutants and haze, which consists of tiny water-coated particles suspended in the air, often obscure vision. Haze frequently occurs during the summer, since it forms most efficiently in warm air. The densest haze can reduce visibility below three miles and cast a yellowish-brown appearance to distant objects. Blowing dust, on the other hand, can reduce visibility to zero. Blowing dust mainly occurs in arid regions of the world, where strong winds sometimes loft large amounts of dust and sand into the air. Dense fog can also limit visibility to near zero. Driving a car in dense fog or blowing dust is extremely dangerous. Many deadly, multi-vehicle pileups on interstate highways have occurred in regions of dense fog or blowing dust.

PRECIPITATION AND VISIBILITY

Falling precipitation, such as rain or snow, also limits visibility. When it is raining or snowing, the visibility often corresponds to the rate at which the precipitation is falling. The heavier the rain or snow, the lower the visibility will be. Snow, in particular, is effective in reducing vision because of its size and opacity.

Meteorologists can tell how heavy snow is falling, and at what rate it is likely accumulating, by the visibility. Snow that obscures visibility to 1/4 mile or less is

In bad snowstorms, visibility can be reduced to dangerous levels.

called heavy snow and can accumulate an inch per hour or more. Moderate snow, accumulating an inch or so every couple of hours, reduces visibility to between 1/2 and 3/4 mile. Light snow, with minor or no accumulation, obscures visibility to 3/4 mile or more. *See also: Air; Fog; Haze; Precipitation; Rain; Snow.*

WARM FRONT

A warm front marks the advancing edge of relatively warmer and more humid air. On a weather map, it appears as a red line with semicircles pointing in the direction the front is moving. A warm front can be hundreds of miles long and extend thousands of feet into the atmosphere. It typically moves more slowly than a cold front, causing widespread cloudiness and steady, mostly light precipitation.

Even though a front is depicted on a weather map as a distinct line, in the atmosphere a front actually is a much wider zone of transition between air of different molecular density. A warm front can be hundreds of miles wide at the earth's surface, with a gradual change in weather from the warmer to the cooler side of the front. Warm fronts do not extend straight upward into the atmosphere; rather, they gradually slope upward in the direction of the front's movement.

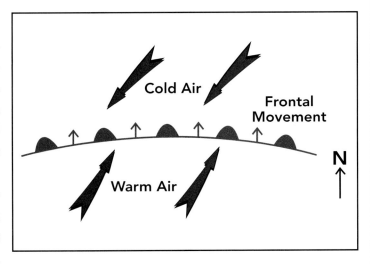

A warm front marks the leading edge of advancing warm air.

The sloping structure to a warm front occurs because warm, moist air is less dense than the cold, dry air. As the warm front advances, the warmer air rides up and over the denser cold air ahead of it. Because advancing warm air overrides colder air ahead of it, rather than wedging or plowing into the cold air, warm fronts move relatively slowly, typically at 15 to 20 mph.

This gradual, sloping structure results in relatively weak upward vertical air currents along and ahead of the front. As air rises, it cools and expands, allowing water vapor in the air to condense into clouds that can produce rain or snow.

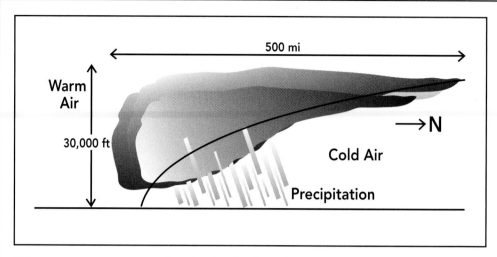

A cross section of a warm front shows how warm air rides up and over colder air at the surface.

This process occurs over hundreds of miles along and ahead of the warm front, causing widespread clouds and steady precipitation. Since vertical air currents are relatively weak along a cold front, the precipitation will usually be light.

Meteorologists analyze weather maps to find the location of fronts. To do this, they look at temperature, moisture, air pressure, and wind readings taken from hundreds of weather stations across the country. This data is plotted on a map, making weather patterns easy to visualize. A shift in wind direction from southerly to northerly across a region can mark a warm front. Another way to find a front on a weather map is to examine air pressure readings. Air pressure usually falls ahead of a warm front, reaches a minimum along the front, then rises behind the front. Once a front has been identified, meteorologists can tell if the front is warm by whether the warmer, more humid air on one side of the front advances over time. Temperatures and humidity levels will be lower ahead of the warm front, usually to the north or east, and higher behind the front to the south or west. **See also: Air; Atmosphere; Cloud Formation; Cold Front; Precipitation; Rain; Snow; Water Vapor; Weather Maps.**

WATERSPOUT

A waterspout is a rotating column of air extending from the base of a cumulus or cumulonimbus cloud to the surface of a body of water. Waterspouts are similar to tornadoes but form under different circumstances and are usually much weaker. They are common over warm ocean water and dissipate shortly after crossing land.

Cumulus or cumulonimbus clouds are large and puffy and can grow to great heights. They form when air near the earth's surface warms, becomes less dense, and rises, carrying moisture high into the atmosphere. The warm, humid air above tropical water at a temperature of 80° or higher is ideal for the rapid growth of cumulus and cumulonimbus clouds. The water quickly warms the air above and supplies abundant moisture, as the warm ocean water evaporates into gaseous water vapor.

A waterspout spins across the surface of the ocean.

Waterspouts form in regions of converging air underneath rapidly growing cumulus or cumulonimbus clouds, as air collides and spins up into a column. Air spreading horizontally at the earth's surface continuously replaces the air rising into the cloud. As air flows inward from all directions underneath the cloud, it collides, or converges. The flatness of the water surface allows air to spin up smoothly into a waterspout. A similar phenomenon, called a landspout, occurs less frequently underneath rapidly growing, large cumulus or cumulonimbus clouds over flat regions of land. In both waterspouts and landspouts, wind speeds typically do not exceed 60 mph. However, this is still enough to capsize a small sailboat and damage larger boats. Waterspouts can also cause minor damage as they move onshore before dissipating.

In the United Sates, waterspouts are most common in the tropical waters around the Florida Keys. However, they can occur over any large body of water. Tornadoes can be briefly classified as waterspouts when they travel over rivers or lakes. A tornado is different from an ordinary waterspout in that it retains its considerable strength when moving over water. *See also: Air; Atmosphere; Cloud Classification; Cloud Formation; Oceans; Temperature; Tornadoes; Water Vapor.*

Water Vapor

Water vapor is the invisible, gaseous form of water. It is always present in the atmosphere in varying amounts, with the highest concentration of water vapor molecules found near the earth's surface. Water vapor is responsible for the formation of clouds and precipitation and contributes to the regulation of the earth's temperature.

Clouds form when water vapor condenses into billions of tiny liquid cloud droplets.

Together with oxygen, nitrogen, and other gases, water vapor is one of the constituents of air. In the troposphere, the lowest layer of the atmosphere, water vapor accounts for up to 4 percent of the molecules in the air. The troposphere extends vertically to a depth of seven miles and contains nearly all of the clouds and precipitation that occur in the atmosphere. Water vapor is most plentiful in the troposphere because large amounts of water vapor are generated by the oceans and other bodies of water at the earth's surface.

CONDENSATION AND EVAPORATION

Water vapor consists of individual water molecules moving fast enough to remain separated. Water vapor changes into liquid water through the process of condensation, while liquid water changes into water vapor through the process of evaporation. Condensation occurs when water vapor molecules combine and stick together, becoming a tiny liquid water droplet. Evaporation occurs when the molecules in water fly apart into individual water vapor molecules. Condensation and evaporation occur continuously and simultaneously in the atmosphere. Water vapor molecules are always colliding and sticking together briefly, then flying apart again.

Meteorologists are concerned with determining if there is net condensation or net evaporation. Net condensation occurs when the rate of condensation exceeds the rate of evaporation. This means that even though some water molecules continue to evaporate into water vapor, more water vapor molecules condense into liquid. Likewise, net evaporation occurs when the rate of evaporation exceeds the rate of condensation. The formation of clouds and fog depends on whether net condensation occurs in the atmosphere. Clouds and fog dissipate when net evaporation occurs.

Condensation and evaporation rates depend on how fast the water vapor and water molecules are moving. The average speed of molecular movement is measured by the air temperature. The higher the temperature, the faster the movement of the molecules, and the more frequently water molecules will break apart, or evaporate, into water vapor molecules. Conversely, the lower the temperature, the slower the molecular movement and the more frequently water vapor molecules will stick together, or condense, into water molecules.

EFFECTS OF WATER VAPOR IN THE ATMOSPHERE

Clouds form when there is a net condensation of water vapor molecules in rising air. As air rises through the atmosphere, it moves through layers of lower and lower air pressure, since less and less of the atmosphere remains to weigh down from above. This allows the air to expand and become less dense. Air molecules, including water vapor, move more slowly in less dense air because the space between the molecules increases. This slower molecular motion

Condensed water vapor formed these mammatus clouds underneath a thunderstorm over Teton National Park in Wyoming.

translates into a lower air temperature. Eventually, the rate at which water vapor molecules collide and stick together exceeds the rate at which water vapor molecules fly apart. Net condensation occurs and millions of tiny water droplets develop, causing a cloud to form.

Water vapor also regulates the atmospheric temperature in two different ways: condensation and evaporation. When water vapor condenses into liquid water, it releases energy into the atmosphere, causing the air temperature to increase. The release of energy from condensation can cause clouds and storms to grow larger and stronger. This is because the energy from condensation causes the air to become warmer than it would otherwise be. Warmer air is less dense than cooler air at the same pressure, and it will rise higher in the atmosphere. This contributes to the growth of a cloud or storm.

When liquid water evaporates into water vapor, on the other hand, it removes energy from the surrounding atmosphere, causing the temperature to fall. One example of the effect of evaporation in the atmosphere occurs when wind blows over mountain ranges. As air moves from a higher to lower elevation, more of the atmosphere exerts pressure on it from above. The air therefore compresses and warms as the air molecules move around faster. Water vapor molecules in air also gain speed and fly apart from one another with greater frequency. The energized water molecules fly apart more often than they stick together, and any liquid water droplets, such as cloud or rain droplets, will tend to evaporate. As moisture evaporates in descending air, clouds and precipitation frequently dissipate downwind of large mountain ranges.

Another way that water vapor regulates air temperature is by absorbing radiation emitted by the earth. Radiation is energy in the form of invisible electromagnetic waves, emitted and absorbed at varying wavelengths by all objects. The earth emits radiation with a relatively long wavelength, and water vapor is an effective absorber of this outgoing "terrestrial" radiation. As water vapor molecules absorb some of the earth's radiation they gain energy. These energized molecules, in turn, emit radiation towards the earth, causing it to become warmer. As it becomes warmer, the earth emits more terrestrial radiation into the atmosphere. This continuous cycle of radiation exchange between the earth and the atmosphere is known as the greenhouse effect. Because of greenhouse gases such as water vapor and carbon dioxide, which effectively absorb outgoing terrestrial radiation, the earth is significantly warmer than it would otherwise be if these gases were absent. *See also: Air; Atmosphere; Cloud Formation; Fog; Greenhouse Effect; Oceans; Precipitation; Radiation; Temperature.*

WEATHER

Weather is the condition of the atmosphere at any given time and location. It includes the temperature, pressure, and humidity of the air; the wind direction and speed; and phenomena such as clouds, rain, and snow. Weather arises through the large-scale horizontal and vertical movements of air. These movements occur because of the uneven heating of the earth by the sun.

There are many factors that cause the earth to become unevenly heated. The first is the orbit of the earth around the sun. The earth moves in a slightly elliptical orbit, rather than a circular one. This means that at some times of the year, the earth is farther away from the sun and receives somewhat less solar energy than other times. More important, the earth is tilted on its axis in a fixed direction. This causes the Southern and Northern Hemispheres to alternately point toward and away from the sun at different times of the year. When one hemisphere points toward the sun, it receives more solar energy. Temperatures rise during this part of the year, known as summer. When the hemisphere points away from the sun, it receives less solar energy, causing temperatures to fall as winter sets in.

In addition to seasonal temperature changes, the temperature varies from north to south because of the shape of the earth. Since the earth is a sphere, the sun's rays strike the earth at different angles depending on latitude. Locations closer to the poles receive less solar energy because the sun's rays strike the earth less directly. Areas nearer the equator receive more solar energy because the sun's rays strike the earth more directly. In general, this causes temperatures to be warmer near the equator and lower near the poles.

The varying terrain of the earth itself also contributes to temperature differences. Land warms and cools at a much faster rate than water. The oceans, therefore, never become as cold nor as warm as land areas. Furthermore, changing elevations result in temperature differences over small distances. High elevations are colder, on average, than low elevations.

Weather occurs because of the relation between temperature, pressure, and wind. Changes in temperature from one location to another result in changes in air pressure, since air pressure is a function of temperature. Air tends to move from areas of higher pressure to areas of lower pressure. Thus, temperature differences yield pressure differences, which cause the wind to blow. Wind is also highly influenced by the motion of the spinning earth. As the earth spins underneath the atmosphere, it imparts motion to it. Air, therefore, moves as a result of the spinning earth and as a result of pressure and temperature differences. Wind transports areas of cold and warm air, moist and dry air across the globe. These winds swirl and mix horizontally, giving rises to vast circulations around areas of low and high air pressure.

While air moves freely around the earth in horizontal directions, it also moves vertically up and down through the atmosphere. In fact, most weather is the result of these vertical air currents. Air does not move as freely in the vertical direction as it does in the horizontal direction largely because of the restraining effects of gravity. Nevertheless, varying patterns of air pressure often permit air to rise through the atmosphere. When this rising air contains abundant moisture, it can cause clouds to form. If the clouds become large enough, they will eventually produce rain or snow.

Rising air, clouds, and precipitation are limited mainly to the lower portion of the atmosphere. This region of the atmosphere is called the troposphere. Above the troposphere exists the stratosphere, a stable layer of air extending up to thirty

Rain falling from a cloud is a result of rising air currents.

miles or so in altitude. The stratosphere is called stable because it prevents air from rising. As air rises up through the troposphere from below, it hits the bottom of the stratosphere and stops rising. Therefore, virtually all the weather that people experience is confined to the lowest seven miles or so of the atmosphere, even though the atmosphere itself extends hundreds of miles up. ***See also: Air; Atmosphere; Cloud Formation; Oceans; Precipitation; Rain; Snow; Temperature; Wind.***

WEATHER EXTREMES

The following tables show record weather extremes by location and date. Weather records are recorded only at official weather observing sites, which are usually established and regulated by a national government. The vast number of recording stations around the world ensures that the weather records listed here are representative of the most extreme weather conditions on the earth.
However, weather extremes that may exceed those officially recorded can and do occur elsewhere. A number of these record weather events occur in countries that do not keep reliable weather records. ***See also: Weather***.

The Macmillan Encyclopedia of Weather

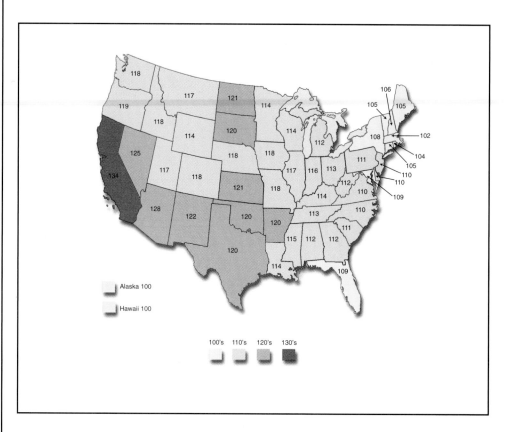

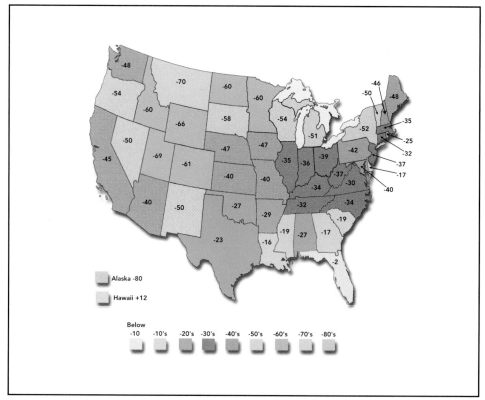

High Temperature Records

Region	Temperature (F)	Date	Location
World	136	September 13, 1922	El Azizia, Libia
North America	134	July 10, 1913	Death Valley, California
Asia	129	June 21, 1942	Tirat Tsvi, Isreal
Australia	128	January 16, 1889	Cloncurry, Queensland
Europe	122	August 4, 1881	Seville, Spain
South America	120	December 11, 1905	Rivadaiva, Argentina
Antarctica	59	January 5, 1974	Vanda Station

Low Temperature Records

Region	Temperature (F)	Date	Location
World	-129	July 21, 1983	Vostok, Antarctica
Asia	-90	February 7, 1892	Verkhoyansk, Russia
North America	-81.4	February 3, 1947	Snag, Yukon, Canada
United States	-79.8	January 23, 1971	Prospect Creek, Alaska
Europe	-67	Unknown	Ust'Shchugor, Russia
South America	-27	June 1, 1907	Sarmiento, Argentina
Africa	-11	February 11, 1935	Ifrane, Morocco
Australia	-9.4	June 24, 1994	Charlotte Pass, NSW

World Record Rainfall Rates

Type	Amount (in.)	Date	Location
Greatest 1-minute	1.23	July 4, 1956	Unionville, Maryland
Greatest 1-hour	12.0	January 22, 1947	Holt, Missouri
		January 24, 1956	Kilauea Sugar Plantation, Hawaii
Greatest 1-day	72	January 7-8, 1966	La R'eunion Island

Greatest Average Annual Precipitation

Region	Amount (in.)	Location
World	467.4	Mawsynram, India
United States	460	Mt. Waialeale, Kaui, Hawaii
South America	354	Quibdo, Columbia
Australia	340	Bellenden Ker, Queensland
North America	256	Henderson Lake, British Columbia, Canada
Europe	183	Crkvice, Bosnia-Herzegovina

Least Average Annual Precipitation

Region	Amount (in.)	Location
World	0.03	Arica, Chile
Africa	less than 0.1	Wadi Halfa, Sudan
North America	1.2	Bataques, Mexico
United States	1.63	Death Valley, California
Asia	1.8	Aden, South Yemen
Australia	4.05	Trovdanina, South Australia
Europe	6.4	Astrakha, Russia

World Record Snowfall Rates

Type	Amount (in.)	Date	Location
North America Greatest 24 hours	76	April 14-15, 1921	Silver Lake, Colorado
Europe Greatest 24 hours	68	April 5-6, 1959	Bessans France, French Alps
World Highest Seasonal Total	1140	1998-1999	Mt. Baker, Washington

World's Largest Hailstone

Circumference (in.)	Weight (lbs)	Date	Location
17.5	1.67	September 3, 1970	Coffeyville, Kansas

Air Pressure

Type	Air Pressure (in.)	Date	Location
World's Highest	32.01	December 31, 1968	Agata, Russia
North America Highest	31.85	January 31, 1989	Northway, Alaska
World's Lowest	25.69	October 12, 1979	Eye of Typhoon Tip, Western Pacific Ocean
North America Lowest	26.35	September 2, 1935	Matecumbe Key, Florida

World's Highest Wind Speed

Speed (mph)	Date	Location
231 mph gust	April, 12, 1934	Mt. Washington, New Hampshire

Weather Forecasting

Meteorologists forecast weather using a combination of techniques and a variety of tools, including computers, satellite photos, and radar. Because of the vast size of the atmosphere and the chaotic motion of air, the farther into the future the prediction is made, the less accurate the weather forecast will be.

Methods of Weather Forecasting

There are many methods for forecasting the weather, and meteorologists typically use a combination of them. One way to forecast weather is called the trend method. In this method, meteorologists analyze the position, movement, and rate of development of weather systems and assume that these trends in the weather pattern will remain consistent over time. For example, a forecaster might predict that a cold front moving eastward at 25 mph will continue to move

Meteorologists are trained in the physics of the atmosphere and learn how fundamental scientific laws govern the motion of air. Knowing how air pressure, density, temperature, and horizontal and vertical currents of air are related, along with scientific theories for the formation of storms, allows meteorologists to understand weather patterns on a deeper level.

Meteorologists in Wallops Island, Virginia use this twenty-six meter satellite antenna to receive images from polar orbiting satellites.

eastward at this same rate of speed in the near future. The trend method is more accurate in very short time periods—on the order of hours rather than days. Similar to the trend method is a persistence forecast. When a meteorologist makes a persistence forecast, he or she predicts that current weather conditions will persist through the forecast period. One reason a forecaster might make a persistence forecast is if weather systems, such as low- or high-pressure areas or fronts, have slowed or stalled, prolonging a particular type of weather condition over a certain area. If cloudy, cold weather is occurring at a given location, for example, a persistence forecast will call for cloudy, cold weather to continue for the next day or so.

Meteorologists can also make predictions based on the visual pattern of weather as displayed on weather maps. Using the analogue method, a weather forecaster can compare and contrast a current weather pattern with one that may have occurred weeks, months, or even years ago. If the two weather patterns are very much alike, the forecaster may predict weather conditions to be similar to the historical analogue weather pattern. An analogue forecast is a type of "pattern recognition" forecast. Meteorologists view dozens of different kinds of weather maps day after day. After years of experience, a forecaster may become aware that certain patterns of weather, as visually depicted on a weather map, correlate with the current weather conditions.

Climatology

No matter which combination of forecasting methods a meteorologist uses, they must have a knowledge of average weather conditions and geography in the area for which they are forecasting. Climatology is the study of the average weather conditions over a period of years. Knowing the climatology of an area will allow the forecaster to frame his or her forecast in relation to average conditions. Forecasting a high temperature, for example, that exceeds the all-time high temperature for a given location by twenty degrees would be a very risky forecast given knowledge of climatology. Familiarity with geography allows meteorologists to understand how mountains, cities, and bodies of water influence local temperature and wind patterns.

Tools Used to Forecast the Weather

Meteorologists use various tools to guide them in making a weather forecast. One tool is radar, which shows the location, intensity, and movement of rain and snow. Satellite photographs show cloud patterns associated with large-scale storm systems and fronts, and allows forecasters to view weather systems over remote or sparsely populated parts of the world. Hourly weather observations from thousands of reporting sites around the world give data on temperature, wind, pressure, humidity, sky cover, and other current weather conditions. This

Weather Modification

Meteorologists use a variety of technical equipment to help them forecast the weather.

observation data is input into computers that create atmospheric models. A model is complex software that runs on a supercomputer and simulates the atmosphere for hours or even days into the future. Because of technological limitations, and because of the size and chaotic nature of the atmosphere, computer models predict weather with varying degrees of accuracy. Forecasters take the computer model as guidance and use their knowledge of weather and forecasting to try to accurately predict the weather. *See also: Atmosphere; Cold Front; High Pressure; Low Pressure; Radar; Rain; Satellite; Snow; Temperature; Weather; Weather Maps; Wind.*

Weather Map

Meteorologists use weather maps to help them visualize weather patterns over time and region. Any aspect of weather, including temperature, pressure, humidity, and wind, can be plotted and analyzed on a weather map. Different kinds of data are combined on maps in different ways, allowing meteorologists to view only the most relevant data in a given weather forecasting situation. While weather maps were once drawn by hand, computers now allow the nearly instantaneous creation of weather maps from raw data.

Forecasters analyze data on weather maps by using data "contours." A contour, also called an isopleth, is a line connecting points of equal weather. Weather data contours are similar to elevation contours that show lines of equal elevation on topographical maps. For example, a weather forecaster can draw contours connecting points of equal temperature. Contours are drawn using a standard interval, or numerical separation between lines. Temperature contours are often drawn with a ten degree interval, showing the fifty degree contour, sixty degree contour, and so on. Computers can rapidly calculate and draw contours for any kind of numerical data set.

Contours allow the forecaster to see, for example, areas of high and low temperature or pressure, areas of high winds, or high humidity levels, not only at the earth's surface but for any horizontal layer of the atmosphere. Since weather systems extend vertically through the atmosphere, meteorologists use weather maps showing weather conditions high in the atmosphere as well as near ground level.
See also: Atmosphere; Isopleth; Temperature; Weather; Weather Forecasting.

Weather Modification

People attempt to modify short-term weather conditions mainly for the purposes of increasing rainfall or dissipating dense fog.

A process called seeding is the primary method of weather modification. Cloud seeding involves spraying microscopic crystals of silver iodide into clouds in order to increase rainfall rates. Silver iodide is used because it serves as an effective ice nuclei, a tiny substance on which ice crystals easily form. Ice crystals are essential in the production of rainfall in clouds because they attract water vapor molecules from cloud droplets.

WEATHER MODIFICATION

Small planes, with canisters of silver iodide attached to their wings, release the chemical as they fly through clouds. Water vapor molecules freeze upon contact with the ice crystal, causing the ice crystal to grow larger. Eventually, ice crystals may grow large enough to descend through the cloud and melt into rain, which then falls to the earth. Because naturally occurring ice nuclei are relatively scarce, silver iodide enables more ice crystals to develop and potentially turn into rain.

The results of cloud seeding are controversial. There is some evidence that it can enhance precipitation rates in certain situations, though its effectiveness is difficult to determine because there is no way to know how much rain a cloud would have produced if it had not been seeded with silver iodide. Cloud seeding is mainly used by farmers in times of dry weather.

Another chemical seeding technique can be used to disperse fog at airports where low visibility can prevent some planes from flying. Planes or helicopters flying over the airport spray tiny particles of dry ice into the fog from above. The dry ice causes the tiny liquid fog droplets to freeze and fall to the earth as a fine dusting of snow. This method, however, works only when fog occurs at temperatures of freezing or less.

Airports use chemical seeding to disperse fog so that planes can take off and land.

In the 1960s, the United States government attempted to seed hurricanes to cause them to weaken. Scientists thought that spraying silver iodide into the outer rain bands around a hurricane would sap strength from the intense hurricane core. The redistribution of energy from the interior to the exterior of the hurricane would, it was hypothesized, lead to an overall weakening of the system. However, the project was eventually abandoned because of inconclusive results. Since then there has been no large-scale effort to modify or weaken weather systems, including those that may pose potential hazards to life and property. The atmospheric processes that drive these weather systems are too complex and vast for chemical seeding, or any other technology, to have much of an effect.

Also, the current level of technology does not permit significant or planned long-term modification of weather. Through the effects of industry, however, humans have inadvertently modified the climate. ***See also: Climate; Fog; Hurricane; Precipitation; Rain; Snow; Water Vapor; Weather.***

WIND

Wind is the collective movement of millions of air molecules. Air moves as a result of differences in air pressure over space, called pressure gradients. The stronger the difference in air pressure, the faster the wind. Meteorologists study wind flow not only near the surface of the earth but also at high levels of the atmosphere. These upper-level winds steer large weather systems across the globe.

THE FORMATION AND DIRECTION OF WIND

The pressure gradients that cause wind form in large part because of variations in temperature, since air temperature is proportional to air pressure and density. These temperature variations generally arise from the uneven heating of the earth by the sun. Therefore, wind can be thought of as a conversion of the sun's energy into moving air molecules. In general, wind blows from areas of high pressure to areas of low pressure. If an area of high pressure is located to the northwest of an area of low pressure, wind would tend to blow from the northwest in the absence of other forces.

However, air does not flow directly from high to low pressure. Another factor that influences wind direction is the Coriolis effect. To an observer fixed on the ground, rotating eastward with the spinning earth, wind appears to curve to the right of its path in the Northern Hemisphere and to the left in the Southern Hemisphere. Air moving from high to low pressure will therefore be deflected to the right. As the air moves closer to the area of low pressure, however, the pressure gradient becomes stronger. The Coriolis effect, which bends the wind to the right, is overcome by the pressure gradient force that pushes air toward low pressure.

The third influence on wind is friction. Wind slows as it comes in contact with the earth. Since the Coriolis effect decreases as wind speed decreases, the rightward-directed pull that the Coriolis effect imparts on wind is further lessened as a result of friction.

The result of these influences is that in the Northern Hemisphere, wind circulates outward from high pressure in a clockwise direction, then spirals counterclockwise in toward low pressure. A rule of thumb states that with the wind at a person's back, lower pressure will be to the left. On a local scale, however, wind direction can change as air flows around buildings, hills, and trees.

MEASURING WIND

Meteorologists measure wind by calculating speed in miles per hour, nautical miles per hour (knots), or meters per second, and by indicating the direction from which the wind is blowing. Wind that changes in a counterclockwise direction over time or space is described as backing, while a clockwise change is a veering wind. The average wind direction over a given period is called the prevailing wind.

Winds typically blow at less than 20 mph near ground level. When a large pressure gradient exists over a relatively short distance, or a strong disturbance develops in the atmosphere, winds can blow much faster. The highest wind speed directly measured on earth was 234 mph recorded on the 6,000 foot summit of Mount Washington, New Hampshire in 1936. This reading was taken from an anemometer, the device used for measuring wind. More recently, scientists probing a violent tornado with Doppler radar near Oklahoma City in May 1999 indirectly measured winds of over 300 mph.

High winds snapped this tree in Southport, Conneticut.

Wind Patterns

Some winds are cyclical. Most people are familiar with an onshore sea breeze that cools the beaches on a hot summer day. This wind, and the offshore land breeze that follows it at night, form as a result of temperature differences between land and ocean. At night, when the land becomes cooler than the adjacent ocean water, a breeze will blow from the higher air pressure over land to the relatively lower air pressure over water. During the day, the sunshine heats the land much faster than the ocean. As the temperature of the air over land rises, the pressure and temperature gradients reverse and a wind blows in from the sea.

Like wind near the earth's surface, winds at high levels of the atmosphere also blow in response to differences in pressure and temperature. These differences exist between the warm tropics and the cold poles, and are a result of the relationship between the density of air and the air pressure at high altitudes in the atmosphere. Colder air near the poles is dense and therefore tends to settle near the surface of the earth. Therefore, far fewer air molecules will exist at high altitudes in a cold, dense atmosphere than in a warm, less dense atmosphere. Since pressure is a measure of the weight of air, air pressures aloft are low over regions of cold air. The opposite is true in the tropics, where the atmosphere is warm. Upper-level winds blow as the result of the difference between lower air pressure aloft at the poles and higher air pressure aloft in the tropics.

Sometimes large areas of cold air gather and spread southward, away from the poles, while warm air spreads northward, away from the tropics. Upper-level winds blow faster as temperature differences increase in response to these spreading and clashing masses of air. These areas of higher wind speeds aloft form a discontinuous band of wind, called the jet stream, that undulates around the globe in response to the moving air masses. *See also: Air; Air Masses; Anemometer; Atmosphere; Coreolis Effect; Doppler Radar; High Pressure; Jet Stream; Low Pressure; Oceans; Temperature.*

WIND CHILL

The wind chill measures how cold the air actually feels as a result of the combination of temperature and wind. The higher the wind, the colder the air feels and the lower the wind chill. The wind chill is used in winter when air temperature becomes low enough and wind becomes high enough to pose a threat to health.

Wind makes the air temperature feel lower by stripping heat away from the body. The faster the wind is blowing, the more quickly the body loses heat. For example, an air temperature of 20°F with a wind chill of -10°F means that even though the actual air temperature is 20°F, the rate of bodily heat loss due to wind is equal to what would be experienced at an air temperature of -10°F with no wind.

Wind chill values well below the actual air temperature indicate that wind is causing a rapid loss of heat from a person's body. Two dangers associated with low wind chill values include frostbite and hypothermia. When frostbite occurs, exposed skin freezes. Extremities such as fingers, toes, ears, and the nose are

Wind Chill Chart

Equivalent Temperature (°F)

Wind Speed (mph) \ Calm	35	30	25	20	15	10	5	0	-5	-10	-15	-20	-25	-30	-35	-40	-45
5	32	27	22	16	11	6	0	-5	-10	-15	-21	-26	-31	-36	-42	-47	-52
10	22	16	10	3	-3	-9	-15	-22	-27	-34	-40	-46	-52	-58	-64	-71	-77
15	16	9	2	-5	-11	-18	-25	-31	-38	-45	-51	-58	-65	-72	-78	-85	-92
20	12	4	-3	-10	-17	-24	-31	-39	-46	-53	-60	-67	-74	-81	-88	-95	-103
25	8	1	-7	-15	-22	-29	-36	-44	-51	-59	-66	-74	-81	-88	-96	-103	-110
30	6	-2	-10	-18	-25	-33	-41	-49	-56	-64	-71	-79	-86	-93	-101	-109	-116
35	4	-4	-12	-20	-27	-35	-43	-52	-58	-67	-74	-82	-89	-97	-105	-113	-120
40	3	-5	-13	-21	-29	-37	-45	-53	-60	-69	-76	-84	-92	-100	-107	-155	-123
45	2	-6	-14	-22	-30	-38	-46	-54	-62	-70	-78	-85	-93	-102	-109	-117	-125

most susceptible to frostbite. Prolonged exposure to low wind chill values may also result in hypothermia, a medical term describing the mental and physical breakdown that occurs when the body temperature drops well below normal.
See also: Air; Frostbite; Hypothermia; Temperature; Wind.

This table shows the resulting wind chill value based on wind speed (far left-hand column) and air temperature (top row).

WIND PROFILER

A wind profiler is a device used for determining wind speed and direction vertically through the atmosphere. It is similar to a Doppler radar, but whereas a Doppler radar detects wind through horizontally-emitted pulses of energy, a wind profiler detects wind through vertically emitted pulses. In so doing, it allows meteorologists to view a cross section of winds thousands of feet into the atmosphere above the wind profiler site.

A wind profiler consists of an array, or grid of small antennae. These antennae send pulses of microwave energy up into the atmosphere. This energy strikes air molecules, cloud droplets, and raindrops or snowflakes and bounces back down to the Doppler sensors on the wind profiler. The frequency at which the microwave energy bounces back down depends on the motion of the particles from which it bounces. If the particles are moving away from the wind profiler, in an upward direction, the microwave energy bounces back at a lower frequency than if the object is moving toward the profiler, or in a downward direction. Wind profilers are also specially attuned to frequency changes that occur when microwave energy bounces off of air molecules with different densities.

Small-scale variations in air molecule density are created by spinning air currents in the atmosphere. A wind profiler can analyze the reflected microwave energy and compute the direction and speed of the wind through a cross section of the atmosphere above the wind profiler site.

Meteorologists use wind profiler data for a couple of purposes. The wind direction can tell the meteorologists whether areas of warm or cold air are spreading over the wind profiler site at different levels of the atmosphere. A wind profiler can show, for example, northwest winds at 25,000 feet in the atmosphere and southerly winds near the earth's surface. A meteorologist might be able to infer from this data that high levels of the atmosphere are becoming colder while low levels of the atmosphere are becoming warmer. When temperature differences increase between low and high altitudes in the atmosphere, the atmosphere can become unstable. An unstable atmosphere is one which allows air to freely rise. Rising air transports moisture aloft, causing clouds and potentially rain or snow to form. Wind profilers, therefore, can assist the meteorologist in determining how weather conditions are changing over the profiler station.

Meteorologists can also use wind profilers to track weather systems. Since weather systems extend vertically through the atmosphere, in addition to extending horizontally across the earth, wind profilers enable forecasters to easily visualize the wind patterns in weather systems as they pass overhead. Cold fronts and warm fronts, in particular, are easily recognizable on wind profilers as the wind direction shifts from south to north from one side of the front to the other. *See also: Air; Atmosphere; Cloud Formation; Cold Front; Doppler Radar; Rain; Snow; Warm Front; Wind.*

WIND SHEAR

Wind shear is the change in wind speed and direction with increasing height in the atmosphere. Wind shear that occurs over large vertical distances in the atmosphere often coincides with the development of different kinds of weather. Wind shear that occurs over relatively small vertical distances can cause turbulence. When the wind shear is particularly strong, the resulting turbulence can pose a hazard to aircraft.

Wind speed and direction always change with increasing height in the atmosphere. Therefore, wind shear always exists. Often the change in wind between low and high levels is relatively small, and the wind shear is described as being weak. Meteorologists and airplane pilots are concerned with large changes in wind speed and/or direction with increasing height. When this happens, wind shear is said to be strong.

Strong wind shear occurring from ground level to high elevations of 20,000 or so feet in the atmosphere are often associated with certain kinds of weather. In the winter, for instance, meteorologists look for veering wind shear when forecasting snow. A veering wind is one that changes direction in a clockwise manner. Snowstorms are typically associated with wind shear that shows veering wind direction with increasing height. Near ground level, winds from the northeast will transport cold air into a region. At low and midlevels of the atmosphere, from 3,000 to 10,000 feet in altitude, winds will veer southeasterly, transporting abundant moisture over the cold air. As wind shear brings cold air and moisture together over the same region, a snowstorm may commence.

Wind shear is also an important ingredient in the formation of thunderstorms. The strongest, longest-lasting thunderstorms form in regions of strong wind shear. Wind from the south near the surface of the earth carries warm, humid air northward. At middle levels of the atmosphere, southwesterly winds transport drier air overhead. This drier air prevents the warm, humid air near the earth's

Wind shear is an important ingredient in the formation of a thunderstorm.

surface from immediately rising into the atmosphere and forming thunderstorms. Instead, the warmth and humidity builds up over hours. When cold air spreads in on westerly winds at high levels of the atmosphere, the temperature difference between the surface of the earth and high altitudes becomes so great that the air rapidly rises through the atmosphere, building a thunderstorm cloud in a very short period of time.

While thunderstorms form in regions of wind shear, thunderstorms can themselves create smaller-scale regions of wind shear that can be very hazardous to pilots. As heavy rain falls out of a thunderstorm cloud, it drags air down with it. This downrushing air spreads out at the earth's surface much like water splashing down from a faucet onto the ground. If the downward current of air is particularly strong, violent churning winds can cause strong wind shear over a localized area as the wind spreads out underneath the thunderstorm. This phenomena is called a downburst. Downbursts led to several air disasters in the 1970s and 1980s before meteorologists and pilots fully understood the phenomenon. Today, all major airports are equipped with wind shear detectors that sense when swirling downburst winds threaten planes near the airport. ***See also: Atmosphere; Downburst; Snow; Temperature; Thunderstorms; Turbulence; Weather; Wind.***

WINTER STORMS

A winter storm is the combination of heavy frozen precipitation, including snow, freezing rain, and sleet, low temperatures, and sometimes high winds, that occur in association with strong areas of low pressure during winter. A winter storm can disrupt travel over thousands of miles and affect millions of people if it occurs over a highly populated region. Winter storms form in regions where masses of air with strongly contrasting temperature coincide with plentiful atmospheric moisture.

DEVELOPMENT OF WINTER STORMS

The areas of low pressure that cause winter storms develop in response to the changing wind patterns and temperatures in upper levels of the atmosphere. An area of low air pressure at the earth's surface, which acts as the center of circulation around a winter storm, develops under a region of strong diverging wind at 15,000 to 25,000 feet aloft. When wind diverges, it spreads apart and removes air from over a region. With less air to weigh down from above, air pressure falls and an area of low pressure develops at the earth's surface underneath an area of diverging wind. Winds often diverge at high altitudes in between large masses of cold and warm air that extend over thousands of miles.

Wind at the surface of the earth spirals in toward the center of low pressure in a counterclockwise direction and rises through the atmosphere to replace the air removed at high altitudes by diverging winds aloft. The circulation around a winter storm can extend for many hundreds of miles, with south-

Winter storms can transform the landscape into a snowy wonderland.

erly winds transporting warm, moist air northward on the south and east side of the low, and northerly winds carrying colder air southward on the north and west side of the low. Warm, moist air rises as it spirals in toward the low. This rising air carries moisture high into the atmosphere, where it condenses, changing into liquid water droplets that form clouds. As more and more air carries greater quantities of moisture upward, the clouds grow bigger and eventually produce precipitation.

On the northern side of the storm, winds circulate toward the low-pressure area from the north and northeast. These cold winds often blow from a large area of high pressure situated hundreds of miles north of the winter storm. Precipitation falling into this cold air falls in the form of snow, sleet, or freezing rain. The heaviest snow and ice occur where the counterclockwise flow of air around the low brings abundant moisture in from the south, up and over the cold air north of the low.

Weather forecasters know that if a cold area of high pressure does not exist north of a developing area of low pressure, the storm will likely be less intense. As diverging winds at high elevations over the winter storm remove more air, the air pressure in the center of the storm continues to fall. As the difference in pressure between the center of the storm and the high-pressure area north of the storm increases, winds around the storm blow faster. These strong winds can cause significant blowing and drifting snow, reducing visibility to mere yards at times in blizzard conditions.

The strong low-pressure areas that produce winter storms can occur anywhere given the right conditions. In general, a winter storm needs a supply of cold air to the north, a supply of warm, moist air to the south, and diverging winds at high elevations overhead. There are other very complex processes that occur in the development of winter storms, including interactions between the atmosphere, land, and oceans. In the United States, the most favored location for strong winter storms is along the eastern seaboard. Here,

Severe winter storms can produce blizzard conditions with dangerously low visibility.

warm ocean currents offshore enhance the temperature contrast between air over the ocean and air over adjacent land areas. When such a weather system approaches this pre-existing region of strongly contrasting temperatures along the East Coast, it can intensify very rapidly. Strong winter storms can also form in the high Plains, often in eastern Colorado, before moving northeast toward the upper Midwest. In this part of the country, air gains spin as it moves over the Rocky Mountains. This extra spin in the atmosphere serves to enhance winter storms as they move out of the Rockies and onto the Plains.

FORECASTING WINTER STORMS

Winter storms are large-scale weather systems that spread a variety of hazardous weather over wide regions. Meteorologists use computers to predict the development of winter storms days in advance. While the accuracy of computer forecasts have been increasing in recent decades, computers have a difficult time resolving small-scale meteorological features associated with winter storms. These features often turn out to be the most important factors in whether or not a winter storm forecast is successful.

Winter storms are best predicted by atmospheric models run on supercomputers. An atmospheric model is a large and complex piece of software that simulates the atmosphere over large sections of the globe for hours or days into the future. Meteorologists study maps of predicted weather patterns generated by these models, as well as numerical output such as temperature and wind forecasts, to guide them in their forecast.

Atmospheric models predict weather conditions at grid points, centered anywhere from 10 km to 50 km or more apart. The grid is placed over a simulated map of the earth, encompassing the region for which the computer model is to predict the weather. Atmospheric models predict not only surface weather conditions but also weather conditions at grids stacked vertically over one another in horizontal layers through the atmosphere. They use extremely complex mathematical equations that compute the motions in the atmosphere based on scientific laws. These equations predict how air changes in temperature, movement, pressure, and moisture levels at each grid point through the simulated three-dimensional model of the atmosphere.

Since the space between each grid point is relatively large, on the order of dozens of miles, atmospheric models cannot see small-scale weather phenomena such as tornadoes and flash floods. Winter storms, however, are very large, extending over thousands of miles. Computer models are therefore able to predict the general large-scale weather patterns that lead to the development of winter storms, sometimes many days in advance. As these models have become increasingly sophisticated over the past few decades, they have also become, on average, more accurate.

Despite increasing accuracy overall, computer predictions of winter storms can sometimes be significantly in error for a couple of reasons. One reason is that the atmosphere is a vast, chaotic system that far exceeds in complexity the current predictive powers of computer models. Another reason is that the equations used by models to predict the motion of air contain approximations and assumptions about how actual atmospheric processes work. These assumptions are necessary to prevent the model from becoming hopelessly complex and taking far too long to process data and compute forecasts. But they also can lead to errors in the forecast output.

Even if the overall model forecast of a winter storm is accurate, the success of a forecast issued by a meteorologist based on the computer output depends on details that the models are often unable to resolve. For example, a computer model may successfully predict the timing, location, and intensity of a large winter storm. One detail of critical importance to the forecaster, however, is the location of the dividing line between rain and snow. This line, called the rain-snow line, marks the boundary between milder temperatures with rain, and colder temperatures with snow. North and west of the rain-snow line, heavy snow or ice can cause life-threatening conditions. South and east of the rain-snow line, people experience nothing more than a nuisance rainstorm. Rain-snow lines typically form on the east side of a winter storm, where mild air circulating north and west around the storm encounters cold air moving south.

The rain-snow line is an atmospheric phenomenon of sufficiently small detail that computer models have difficulty predicting, even if they have accurately predicted the overall storm. To make matters more difficult for the forecaster, the rain-snow line sometimes wavers back and forth over large metropolitan areas. A shift of several miles in one direction or the other may turn an accurate forecast into an inaccurate one, even though the large-scale storm has been successfully predicted. Meteorologists must therefore closely monitor hourly weather observations at dozens of reporting sites in the path of the storm to see exactly how small-scale atmospheric features will affect the severity of the weather at different locations.

THE "SUPERSTORM" OF MARCH 1993

On March 12–15, 1993, a large and powerful winter storm swept up the east coast of the United States, producing record snowfall, blizzard conditions, tornadoes, and coastal flooding. Because of its severity and of the unusually large area affected, it was later referred to both as the "Superstorm" and the "Storm of the Century." Damage estimates from this particularly intense winter storm range

Winter Storms

Heavy snow results in dangerous driving conditions.

from $1.6 to $6 billion with over 300 storm-related fatalities.

While most winter storms that strike the eastern United States form along the Middle Atlantic coast, this winter storm developed unusually far to the south over the warm, tropical waters of the Gulf of Mexico. The storm formed along a zone of strongly contrasting temperatures, where cold, winter air moving southward from the central United States met with warm tropical air over the northern Gulf. Strong wind energy at high levels of the atmosphere, from 15,000 to 30,000 feet in altitude, spread eastward from Texas and interacted with the strong temperature

Even walking is difficult in a winter storm.

contrast to cause air pressures to fall extremely rapidly in the northern Gulf.

The rapidly intensifying low-pressure area moved northeast from the Gulf of Mexico, crossing land over the Florida Panhandle. To the south, a squall line of thunderstorms spawned by the developing winter storm raced across Florida, producing fifteen tornadoes. Strong onshore winds along Florida's west coast drove an unusually high surge of Gulf ocean water inland, causing unexpected and major flooding of low-lying areas. In some locations, the water level along the coast, driven by gale force winds, quickly rose to twelve feet above normal. At least seven people died in the flood waters, while in other parts of the state of Florida, thirty-seven other winter storm-related deaths were reported. Up to six inches of snow fell in the Florida Panhandle, an extremely unusual event in such a warm climate.

As the winter storm moved northeastward along the East Coast, it continued to intensify. Heavy snow fell west of the storm's track, with heavy snowfall and blizzard conditions occurring in parts of the country not accustomed to snow. Parts of Alabama, northern Georgia, and eastern Tennessee were buried under as much as twenty inches of snow, with snowfall records broken in many locations. Up the Appalachians, from North Carolina to northern New England, most locations received nearly two feet of snow. Hundreds of roofs collapsed under the weight of the snow, and electricity was knocked out for thousands of people. All major airports up and down the East Coast were closed, as were all interstate highways.

Along the coast, hurricane force winds of 75–90 mph brought battering waves and flooding along the beaches. Hundreds of homes were damaged, and at least eighteen houses fell into the sea along the Long Island shore. At sea, the Coast Guard rescued 160 people from sinking or damaged vessels. South of Nova Scotia, a 177-meter ship sank in sixty-five-foot seas, with all thirty-three crew members lost. As the Superstorm departed northeast into Canada, bitterly cold air swept down into the eastern United States on northerly winds, setting new record low temperatures in many locations. *See also: Air; Atmosphere; Climate; Cloud Formation; Floods; Freezing Rain; High Pressure; Low Pressure; Oceans; Precipitation; Sleet; Snow; Temperature; Thunderstorms; Tornadoes; Visibility; Weather; Wind.*

ZONAL FLOW

The term "zonal flow" describes the flow of the jet stream in upper levels of the atmosphere. The jet stream is a discontinuous current of fast-moving air at high altitudes, typically 20,000 to 30,000 feet in the atmosphere. It generally flows from west to east but can meander north and south in curves and bends that extend for thousands of miles. In general, a straighter, west to east flow is said to be zonal while a highly curved jet stream flow is described as meridional. When the upper-level air flow is zonal, areas of low and high pressure at the earth's surface are likely to be much weaker.

The jet stream can be likened to the flow of water in a river. Sometimes rivers flow nearly straight, while other times they snake and curve in wide loops back and forth across the land. Zonal flow is similar to a river that flows in a nearly straight line. The direction and speed of jet stream winds are determined by the movement of large masses of cold and warm air around the earth. Jet stream winds blow in between these air masses, in the regions of greatest horizontal temperature contrast. The greater the temperature contrast, the stronger the wind. The direction of air flow is determined by the position of the air masses relative to one another. Jet stream winds blow southward around a large cold air mass, then back northward around a large warm air mass.

Zonal flow is associated with weak areas of low and high pressure at the earth's surface. These low and high pressure areas form underneath regions of diverging and converging winds, respectively. Diverging and converging wind flow removes or adds air over a region, thereby affecting air pressure at the earth's surface. Strong divergence, for example, can remove air from over a region and cause an area of low pressure to form or strengthen near ground level. When the upper-level air flow is zonal, there are few pronounced regions of divergence and convergence. Areas of low- and high-pressure, therefore, remain much weaker on average. *See also: Air; Air Masses; Atmosphere; Convergence; Divergence; High Pressure; Jet Stream; Low Pressure; Temperature; Wind.*

Glossary

Acid Rain: Rain containing acids formed when industrial pollutants released into the air mix with clouds and rain droplets. Acid rain can damage trees, wildlife, and buildings.

Adiabatic Process: The process by which air parcels rise and fall in the atmosphere without exchanging energy with the outside air; the air outside the parcel does not mix with air inside the parcel.

Advection: Changes in atmospheric properties, such as temperature or moisture, caused by the passage of wind across the earth's surface.

Air Mass: A large expanse of air with a relatively uniform density of air molecules. It is usually accompanied by a characteristic type of weather.

Air Mass Weather: Weather that generally accompanies a given type of air mass, which depends on the temperature and moisture of the air mass.

Air Parcel: An imaginary bubble, or volume, of air.

Air: A mixture of gases, microscopic particles, pollutants, and dust, consisting primarily of nitrogen and oxygen, as well as water vapor, carbon dioxide, and ozone.

Albedo: The albedo of an object is the percentage of electromagnetic energy that the object reflects as compared to the amount that strikes it. The higher an object's albedo, the more energy it reflects.

Alberta Clipper: A fast-moving, small area of low pressure that moves through parts of the northern United States, named after the western Canadian province where these storms originate.

Anemometer: An instrument that measures wind speed.

Anticyclone: A high-pressure area.

Aristotle: (384-322 B.C.E.) Greek philosopher, author of *Meteorologica*, the first known book to speculate on the nature of weather, which makes both wildly erroneous and remarkably perceptive hypotheses.

Atmosphere: A layer of gases and microscopic particles surrounding the earth's surface that extends up for several hundred miles to an indefinite threshold with space.

Aurora: A luminous display of color, often shimmering or glowing in appearance, in the upper atmosphere. It is visible at night and occurs most often at high latitudes. In the Northern Hemisphere, it is known as the aurora borealis; in the Southern Hemisphere, it is called the aurora australis. It is caused when energized particles from the sun interact with atmospheric gases.

Barometer: An instrument used to measure the atmospheric pressure.

Blocking Pattern: An upper-level weather system that stalls, thereby blocking the normal west to east progression of weather systems over the middle and high latitudes.

Bora: Fall wind that blows down from the mountains of Bosnia.

Butterfly Effect: Name given to the principle that the flapping of a butterfly's wings in China can cause a tornado in Texas weeks later, or rather that a very small event can have major and unpredictable effects after a long enough time. This means that it is impossible to predict the weather accurately far in advance.

Carbon Monoxide (CO): Colorless, odorless, gas that is a result of the incomplete burning of fossil fuels such as oil, gas, and coal; an atmospheric pollutant and health hazard.

Castellanus: Descriptive term for an altocumulus cloud with a towerlike appearance.

Catalina Eddy: A small-scale circulation in the atmosphere that forms along the southern California coast, named after a small island in the Pacific west of Los Angeles.

Chinook: A foehn wind that blows down from the Rocky Mountains onto the high plains from Montana to Colorado, often bringing rapid thaws and snowmelt.

Chlorofluorocarbons (CFCs): By-products of refrigeration, aerosol propellants, and Styrofoam that play a devastating role in the destruction of ozone in the stratosphere.

Clear Air Turbulence: Turbulence at high altitudes that occurs when the air is clear.

Climate: The long-term average of weather conditions at a given location.

Climate Controls: The factors that influence the prevailing weather in an area over a period of years.

Climate Zones: Large regions that share a similar climate.

Climatology: The study of the long-term average of weather conditions in a given region over a period of years.

Cloud Classification: A descriptive Latin term given to a cloud according to the altitude of its base and its height.

Cloud Formation: Method by which clouds form in the sky when the invisible water vapor in air condenses into tiny, visible water droplets or freezes into tiny ice crystals.

Cloud Seeding: Spraying microscopic crystals of silver iodide into clouds in order to increase rainfall rates.

Cold Front: The advancing edge of a relatively colder and drier mass of air.

Cold Wave: A prolonged period of temperatures well below the average for the season.

Condensation: The change of water vapor molecules into droplets of liquid water, usually caused by falling temperatures.

Glossary

Conduction: Transfer of heat energy via collisions between the more energetic molecules of a hotter object and the slower moving molecules of a colder object.

Convection: Transfer of heat energy via the movement of large volumes of a gas or liquid.

Convergence: The condition of the flow of air in a region of the atmosphere when more air enters the area than is able to exit.

Coriolis Effect: The Coriolis effect or force is the phenomenon whereby the path of a moving object appears to bend toward the right in the Northern Hemisphere and to the left in the Southern Hemisphere. It is caused by the earth's rotation and acts on objects not connected with the earth, such as moving air, airplanes, and water.

Cumulonimbus: A very large cloud, extending from several hundred to a thousand or so feet off the ground to as high as 50,000 feet in the atmosphere, and covering an area of perhaps fifty square miles.

Cyclogenesis: The formation and strengthening of a low-pressure area.

Cyclone: An area of low pressure.

Degree Day: A numerical measurement of how warm or cold a season has been; not a unit of time, but rather of temperature.

Derecho: A fast-moving, long-lived group of thunderstorms that produces very high winds along its path.

Desertification: The process whereby semiarid regions become desert.

Dew: Tiny water droplets that form on exposed outdoor surfaces when air cools.

Dew Point: A measure of how much water vapor is contained in a volume of air; the temperature to which air must be cooled for the rate of condensation to exceed the rate of evaporation.

Divergence: The condition of the flow of air in a region of the atmosphere when more air exits the area than is able to enter.

Doppler Effect: The relationship between the frequency at which energy bounces off or is emitted from an object and the object's speed. The frequency changes depending on the energy's direction of movement. A practical illustration of this effect is the change in pitch of the noise a car makes as it moves toward and then past an observer.

Doppler Radar: A type of radar that in addition to sensing precipitation intensity and location, also detects wind direction and velocity.

Downburst: A violent rush of air about five miles in diameter that exits the base of a thunderstorm and spreads out upon hitting the earth.

Drought: A period of abnormally dry weather, lasting months or even a year or more.

Dryline: The atmospheric boundary that forms when moist tropical air meets up with dry continental air.

Dust Storm: A storm that forms when strong winds pick up loose dust and sand from the ground and carry it into the air, severely reducing visibility.

Easterly Wave: An atmospheric disturbance over tropical regions, so-called since the air flows from east to west in the tropics. Occasionally, an easterly wave can grow into a hurricane as it moves over warm, tropical ocean water. Also called a tropical wave.

Eddy: A small-scale spiraling current of air, similar to the small swirling currents of water seen in rivers near rocks.

Electromagnetic energy: *see* Radiation.

El Niño: Spanish for boy child; an occasional warming of sea surface temperatures in the equatorial Pacific off the northern coast of South America. El Niño is a common occurrence each year after Christmas and was named in reference to the Christ child.

Environmental Lapse Rate: The change in the atmosphere's temperature with height.

Equinox: The day of the year during which the sun's rays strike the earth directly at a 90° angle along the equator, marking the beginning of spring or fall.

Evaporation: Conversion of liquid water to water vapor, usually as a result of an increase in temperature.

Extratropical Cyclone: A cyclone that forms outside the tropics.

Fall Wind: Type of local wind that falls down mountainsides.

Flash Flood: Rapid and unexpected rise in water above normal levels.

Flood: Event when a body of water overflows its natural boundaries.

Foehn Wind: Local wind that blows from higher to lower elevations, but differs from fall winds in that they are typically warming winds.

Fog: Billions of tiny water droplets suspended in air; essentially, a cloud at the earth's surface.

Fog Seeding: Using dry ice to disperse fog at airports; the dry ice causes the water droplets in the fog to freeze and fall to the earth as a fine dusting of snow.

Fractus: Descriptive term for a cloud with a ragged, fractured appearance.

Franklin, Benjamin: (1706-1790) American inventor, statesman, scientist, and writer. In addition to his many other accomplishments, he was the foremost weather researcher and observer in colonial times.

Freezing Rain: A form of precipitation that occurs during colder months, consisting of falling water droplets that freeze on impact with the ground.

Frost: Tiny ice crystals that form on exposed outdoor surfaces. Frost usually forms during colder months as overnight temperatures drop near or below freezing.

Frostbite: Freezing of skin and underlying body tissue as a result of exposure to extremely cold air.

Glossary

Fujita, Theodore: (1920-1998) Professor of meteorology at the University of Chicago and pioneer in the research of severe local storms. He made important discoveries about thunderstorms, tornadoes, and downbursts.

Fujita Scale: Scale devised by Theodore Fujita rating the destructive capacity of a tornado and its wind speed from zero to five.

Funnel Cloud: The visible portion of a tornado, usually a whirling inverted cone or cylinder.

General Circulation: The flow of air around the earth averaged over a long period of time.

Global Warming: A gradual increase in the average temperature of the earth.

Graupel: A frozen form of precipitation that appears as a small icy or snowlike pellet. While it usually does not fall to the earth in large quantities, it plays a role in the formation of rain, snow, and hail inside clouds.

Greenhouse Effect: The process by which certain atmospheric gases warm the planet by absorbing and retaining outgoing energy from the earth's surface.

Greenhouse Gas: A gas, such as carbon dioxide, that is an efficient absorber of outgoing energy from the earth's surface and aids the greenhouse effect.

Hail: Chunks of ice that fall during a thunderstorm; can vary from the size of large seeds to the size of grapefruits. Not to be confused with sleet.

Halo: A luminous ring circling the sun or moon, produced when light interacts with tiny ice crystals in the atmosphere.

Hazardous Air Pollutants (HAPs): Dangerous chemicals released into the atmosphere, often by accident, as in a chemical spill.

Haze: Microscopic particles suspended in air that reduce visibility and give a pale white or yellow, fuzzy appearance to the sky.

Heat Exhaustion: Illness caused by the buildup of excess heat in the body. Symptoms include elevated body temperature, faintness or nausea, rapid heartbeat, and low blood pressure.

Heat Index: Also known as the apparent temperature; the combination of temperature and humidity to measure how warm the air actually feels.

Heat Stroke: Serious illness caused by buildup of excess heat in the body, occurring after a person has already passed through the heat exhaustion stage. Symptoms include a temperature of 105° F or greater, hot dry skin, rapid heartbeat and irregular blood pressure, rapid and shallow breathing, and mental confusion.

Heat Transfer: Transfer of energy from a hotter object to a cooler object by one of three means: conduction, convection, and radiation.

Heat Wave: A prolonged period of temperatures well above average for the season.

High Pressure: A large atmospheric circulation, the center of which has a higher atmospheric pressure than all surrounding areas. Also called an anticyclone or "high."

Humidity: The amount of water vapor in the air.

Humilis: Descriptive term for a cumulus cloud with a vertical development.

Hurricane: A giant atmospheric circulation that forms over warm tropical ocean water, with a calm, clear eye at the center of the system and winds of 74 mph (65 knots) or higher.

Hydrocarbons: Organic compounds, such as methane, often released by gasoline or industrial chemicals.

Hydrological Cycle: The process by which liquid and gaseous water circulate among the atmosphere, land, and ocean.

Hydrology: The study of the natural flow of water.

Hydrostatic Equilibrium: The balance between forces that act in vertical directions in the atmosphere—the force exerted by the vertical difference in air pressure is counterbalanced by the force of gravity.

Hygrometer: A device used to measure the humidity of the air.

Hypothermia: A dangerous and life-threatening health condition that occurs when the body temperature drops below 95°F.

Ice Age: A period in the earth's history characterized by low temperatures and expansion of glaciers and polar ice caps.

Intertropical Convergence Zone (ITCZ): A belt of scattered cloudiness, showers, and thunderstorms that stretches around the planet near the equator.

Inversion: A layer of air, often hundreds or thousands of feet thick, in which the air becomes warmer as the altitude increases, the opposite of the normal condition of air.

Ionosphere: Layer of the atmosphere spanning the thermosphere and the mesosphere that reflects shortwave and AM transmissions back toward the earth.

Isallobar: Isopleth used to map air pressure change.

Isobar: Isopleth used to map air pressure.

Isodrosotherm: Isopleth used to map dew point.

Isohyet: Isopleth used to map rainfall or snowfall.

Isopleth: Lines that connect equal data points on a weather map.

Isotach: Isopleth used to map wind speed.

Isotherm: Isopleth used to map temperature.

Jet Stream: A discontinuous current of air that flows generally from west to east around the globe at high altitudes in the atmosphere.

Glossary

Köppen System: System for classifying climate zones into five major categories.

Lake-Effect Snow: Localized, and sometimes heavy, snow that develops downwind of exposed lake water.

La Niña: Spanish for the girl child; an area of cooler-than-average ocean water in the tropical eastern Pacific off the coast of South America; the counterpart of El Niño.

Lenticularis: Descriptive term for a cloud with a lenslike appearance.

Lightning: A powerful electrical discharge originating inside a thunderstorm.

Local Winds: Winds that recur in certain locations under similar circumstances.

Low Level Jet: A region of fast-flowing air in low levels of the atmosphere, generally located anywhere from several hundred to 5,000 feet in altitude and extending several hundred miles in length, with wind speeds generally from 30 to 60 mph.

Low Pressure: A large atmospheric circulation, the center of which has a lower atmospheric pressure than all surrounding areas. Also called a cyclone or "low."

Meridional Flow: Descriptive term for the highly curved flow of the jet stream in upper levels of the atmosphere.

Mesocyclone: A large, rotating column of air inside a thunderstorm that can lead to the formation of a tornado.

Mesoscale: Scale of motion describing weather systems ranging in size from a mile to approximately 100 miles and lasting up to several hours.

Mesoscale Convective Complex (MCC): A large conglomeration of thunderstorms that typically develops and reaches peak intensity during the nighttime hours.

Mesosphere: The next-to-last layer of the atmosphere, beginning over fifty miles from the earth's surface.

Meteorology: The study of the earth's atmosphere, encompassing the observation and forecasting of weather over short time periods, as well as the study of weather over long time periods and the interaction between atmosphere and ocean.

Microclimate: A small-scale region in which average weather conditions are measurably different from a larger, surrounding region.

Microscale: Smallest scale of motion, describing weather phenomena occurring over distances from a few inches to a mile, and lasting for a period of seconds to minutes.

Mie Scattering: Scattering caused by relatively large objects, such as water droplets and dust particles, that causes the scattered light to appear white. Mie scattering is why clouds have a whitish appearance.

Mirage: A false image that appears when light reflected from an object changes path after passing through air of varying density, causing the object to appear above or below its actual position.

Mistral: Fall wind that blows down the Rhône valley in France.

Monsoon: Seasonal shifts in wind direction that occur in parts of the world, best known as a wet, warm-season wind carrying drenching rains. It can also be used to describe the wintertime wind shift that carries dry, cooler air.

National Weather Service: A government agency that issues weather forecasts and warnings to the public twenty-four hours a day. One part of a larger organization called the National Oceanic and Atmospheric Administration (NOAA), which exists as part of the U.S. Department of Commerce.

Nitrogen Oxides: Gasses such as NO2, NO3, and others, created by the burning of fossil fuels and industrial emissions, that contribute to smog and acid rain.

Nor'easter: A strong area of low pressure that moves northward along the United States' Eastern Seaboard. The name is derived from the strong, northeasterly gales that blow in advance of the storm.

Numerical Weather Prediction: The process whereby computers use complex mathematical equations to predict the future state of the atmosphere given data that describes the initial state of the atmosphere.

Occluded Front: A boundary in the atmosphere that marks the division between two relatively cold masses of air.

Ocean: An enormous body of salt water that covers millions of square miles.

Omega Block: Blocking pattern where an upper-level wind takes on the horseshoe shape of the Greek letter omega (Ω). Unlike a Rex block, the jet stream simply diverts around the block, rather than splitting into branches.

Orographic Lift: Term referring to how air is forced to rise as it blows across sloping terrain or against mountain ranges.

Overrunning: The process whereby relatively warm air flows up and over colder air near the earth's surface.

Ozone Hole: A significant depletion of the ozone layer over Antarctica that was discovered in the 1980s.

Ozone Layer: A layer of the stratosphere where ozone absorbs incoming ultraviolet (UV) radiation from the sun, preventing it from reaching the earth's surface.

Ozone (O_3): A gas, the molecules of which consist of three oxygen atoms.

Paleoclimatology: Study of the climate, especially the atmosphere, of the earth thousands and even millions of years ago.

Particulate Matter: Microscopic particles suspended in the air, which can include smoke, dust, and vehicle emissions, and which is the major cause of haze.

Planetary Scale: Largest scale of motion, describing events that exist over thousands of miles, lasting for months or years.

Plate Tectonics: The action of the slow movement of large sections, called plates, of the earth's crust.

Glossary

Polar Front: Discontinuous boundary that undulates around the earth at the middle to upper latitudes and separates the warmer air that originates from tropics and the colder air that originates near the poles.

Polar Vortex: A belt of high winds that forms around Antarctica each winter, sealing the Antarctic atmosphere off from the rest of the planet.

Pollution: Unhealthy or damaging human-made and natural molecules in the atmosphere.

Precipitation: Any form of water or ice, such as rain or snow, that falls from the sky.

Pressure Gradient Force: Force that tends to move air from areas of higher pressure to areas of lower pressure.

Pressure Gradients: Differences in air pressure over space.

Radar: Acronym for radio detection and ranging. In meteorology, radar is used to detect the location, intensity, and movement of precipitation such as rain or snow.

Radiation: Energy in the form of electromagnetic waves, invisible waves of energy that travel at the speed of light. Unlike sound waves or water waves, they do not need a substance, or medium, through which to travel.

Radiosonde: A box of instruments attached to a weather balloon, used to get information from high in the atmosphere.

Rain: Liquid water drops that fall from a cloud to the ground.

Rain Shadow: An area of relatively low rainfall that lies downwind of mountain ranges.

Rainbow: A luminous, brightly colored arc that appears in the sky when sunlight bounces off of raindrops.

Rayleigh Scattering: Scattering caused by tiny air molecules, resulting in the scattering of blue light. Rayleigh scattering is why the sky appears blue.

Relative Humidity: The ratio of water vapor in the air to the maximum amount of water vapor that air can contain at that temperature. As temperature increases, it can contain more water; therefore, the same amount of water vapor in the air at two different temperatures can result in two different relative humidity readings. Because of this, meteorologists prefer to use the dew point as a measure of how much water vapor is in the air.

Rex Block: A blocking pattern where the jet stream encounters a stalled ridge and splits into two branches.

Ridge: An area of relatively high air pressure.

Rossby, Carl-Gustav: (1898-1957) Swedish-American meteorologist whose work led to major advances in the atmospheric sciences in the twentieth century.

Santa Ana: Foehn wind that blows in southern California.

Satellites: Orbiting machines used to take pictures of the earth from high above. These pictures give valuable information on the location and movement of weather systems.

Scales of Motion: Categorization of weather systems using scales describing phenomena occurring over a given time and area.

Scattering: Deflection of sunlight by extremely small objects in the atmosphere; the reason the sky is blue, clouds are often seen as white, and the sun frequently appears red or orange when it sets and rises.

Sea Breeze: Type of local wind that occurs at the seashore.

Seasons: Naturally occurring changes in weather and cycles of plant growth that occur as the earth orbits the sun, caused by the changes in the amount of sunlight that reaches the earth.

Semiarid: Regions of the planet that receive relatively low amounts of annual rainfall and support the growth of not much more than short, drought-resistant grasses.

Sleet: Form of precipitation that occurs during colder months, consisting of falling ice pellets, the size of large seeds, which ping and bounce on impact. Sometimes confused with hail.

Smog: A fog or dark brown haze in the atmosphere caused by industrial pollutants that can cause respiratory illness in humans.

Snow: Form of precipitation consisting of ice crystals that fall from a cloud to the ground.

Snowflake: Special kind of ice crystal that develops tiny, intricate branches as it grows; it can form in a cloud when enough moisture is present and the temperature at cloud level is below freezing.

Solar Flare: A huge eruption of gas from the sun, which can disrupt radio and satellite communication by sending large amounts of particles into the earth's upper atmosphere; this can also result in an aurora becoming visible.

Solstice: The points on the earth's orbit around the sun in which a hemisphere of the earth receives the most (summer solstice) and the fewest (winter solstice) hours of daylight and that marks the beginning of summer or winter.

Sounding: An observation, or set of observations, of weather conditions taken vertically through the atmosphere.

Stability: A measure of the ease with which air can rise. A stable atmosphere is one that prevents air from rising, while an unstable atmosphere allows air to rise freely.

Stationary Front: An atmospheric boundary, moving very little or not at all, that separates warmer air from colder air.

Storm Chasing: Use of vehicles to track and intercept severe thunderstorms and tornadoes, in order to gather data used in the scientific research of severe storms and to provide visual observations that confirm indications of severe storms from remote radar stations.

Storm Surge: A term used to describe the rise in ocean water levels that occurs along a coast as a hurricane moves onshore. It is caused when strong onshore-directed wind piles up water along the coastline.

Glossary

Stratosphere: Layer of the atmosphere extending from seven to thirty miles above the earth's surface, lying between the troposphere and the mesosphere.

Sulfur Dioxide: Atmospheric pollutant produced by the burning of fossil fuels that causes smog and acid rain.

Sun Dogs: Two luminous spots on either side of the sun that occasionally appear as a result of sunlight interacting with ice crystals in the atmosphere.

Sunspot: A cooler area of the sun's surface, surrounded by extremely bright rings, which increase in frequency over an eleven-year cycle; at the peak of the cycle, solar flares are more common.

Supercell: A violent, long-lasting thunderstorm that contains a mesocyclone.

Synoptic Scale: Scale of motion describing systems that extend for a hundred to a thousand or so miles in diameter and last for hours or days.

Tangent Arc: A bright spot on a halo appearing directly above the sun or moon.

Temperature: A measure of an object's average molecular energy.

Thermometer: A device used to measure temperature.

Thermosphere: Outermost layer of the atmosphere, extending to almost 600 miles above the earth's surface.

Thunderstorms: The combination of rain, lightning, and thunder produced by a cumulonimbus cloud.

Tornadoes: A violently rotating column of air extending from the base of a thunderstorm cloud to the surface of the earth.

Trade Winds: Tropical easterly winds, so named because trading ships used them centuries ago to travel from east to west across the oceans.

Tropical Wave: *See* Easterly Wave.

Troposphere: Lowest layer of the atmosphere, extending from the earth's surface to an altitude of approximately seven miles.

Trough: An area of relatively low air pressure.

Turbulence: The chaotic, swirling motion of air over a relatively small distance. It is produced when wind interacts with fixed obstacles, or when wind speed and direction rapidly changes with increasing height in the atmosphere.

Updraft: A current of air that rises to the center of a thunderstorm.

Upper-Level Ridge: A large bend in the flow of air in the upper levels of the atmosphere, moving first toward the pole then back toward the equator.

Upper-Level Trough: A large bend in the flow of air in the upper levels of the atmosphere, moving first toward the equator then back toward the pole.

Urban Heat Island: An area of relatively warm air generated by and centered over a city.

Visibility: A measure of the maximum distance at which a person can see objects clearly.

Warm Front: The advancing edge of relatively warmer and more humid air.

Waterspout: A rotating column of air extending from the base of a cumulus or cumulonimbus cloud to the surface of a body of water; similar to a tornado but formed under different circumstances and usually much weaker.

Water Vapor: The invisible, gaseous form of water.

Weather: The condition of the atmosphere at any given time and location, including the temperature, pressure, and humidity of the air; the wind direction and speed; and phenomena such as clouds, rain, and snow.

Weather Forecasting: Use of a combination of techniques and a variety of tools, including computers, satellite photos, and radar to predict what the weather will be in the future. Because of the vast size of the atmosphere and the chaotic motion of air, the farther into the future the prediction is made, the less accurate the weather forecast will be.

Weather Maps: Maps with symbols describing weather conditions; used to help visualize weather patterns over time and region.

Weather Modification: Attempts to change short-term weather conditions, mainly for the purposes of increasing rainfall or dissipating dense fog.

Wind: The collective movement of millions of air molecules. Air moves as a result of differences in air pressure over space; the stronger the difference in air pressure, the faster the wind.

Wind Chill: Measure of how cold the air actually feels as a result of the combination of temperature and wind.

Wind Profiler: A device used for determining wind speed and direction vertically through the atmosphere.

Wind Shear: The change in wind speed and direction with increasing height in the atmosphere.

Winter Storms: The combination of heavy frozen precipitation, including snow, freezing rain, and sleet, low temperatures, and sometimes high winds, that occur in association with strong areas of low pressure during winter.

Zonal Flow: Descriptive term for the flow of the jet stream in upper levels of the atmosphere when it moves in a straight, west to east direction.

Bibliography

Allaby, Michael. *How the Weather Works.* New York: Reader's Digest, 1995.
Arnold, Caroline. *El Niño: Stormy Weather for People and Wildlife.* New York: Clarion Books, 1998.
Aronsky, Jim. *Crinkleroot's Nature Almanac.* New York: Simon & Schuster, 1999.
Barnes-Svarny, Patricia L. *Skies of Fury: Weather Weirdness Around the World.* New York: Touchstone Books/Simon & Schuster, 1999.
Bilger, Burkhard. *Global Warming.* New York: Chelsea House Publishers, 1992.
Burroughs, William J., et al. *Nature Company Guides: Weather.* Alexandria, VA: Time-Life Books, 1996.
DeMillo, Rob. *How Weather Works.* Emeryville, CA: Ziff-Davis Press, 1994.
De Witt, Lynda. *What Will the Weather Be?* New York: HarperCollins Publishers, 1991.
Discovery Channel Weather: An Explore Your World Handbook. New York: Discovery Books, 1999.
Elsom, Derek. *Weather Explained: A Beginner's Guide to the Elements.* New York: Henry Holt, 1997.
Englebert, Ralph. *The Complete Weather Resource.* Farmington Hills, MI: Gale Group, 1997.
Fishman, Jack. *The Weather Revolution: Innovations and Imminent Breakthroughs in Accurate Forecasting.* New York: Plenum Press, 1994.
Ganeri, Anita. *The Oceans Atlas.* London: Doring Kindersley, 1994.
Greenberg, Keith Elliot. *Stormchaser: Into the Eye of a Hurricane.* Woodbridge, CT: Blackbirch, 1998.
Hardy, Ralph. *Weather.* Lincolnwood, IL: NTC Publishing Group, 1996.
Hodgson, Michael. *Weather Forecasting,* 2d ed. Old Saybrook, CT: Globe Pequot Press, 1999.
Kahl, Jonathan. *Wet Weather: Rain Showers and Snowfall.* Minneapolis, MN: Lerner Publications, 1992.
Knapp, Brian J. *Flood.* Austin, TX: Steck-Vaughn Library, 1990.
Laskin, David. *Braving the Elements: The Stormy History of American Weather.* New York: Doubleday, 1996.
Longshore, David. *Encyclopedia of Hurricanes, Typhoons, and Cyclones.* New York: Facts on File, 1998.
Monmonier, Mark S. *Air Apparent: How Meteorologists Learned to Map, Predict, and Dramatize Weather.* Chicago, IL: University of Chicago Press, 1999.
Morgan, Sally. *Acid Rain.* New York: Franklin Watts, 1999.
Murphree, Tom. *Watching Weather.* New York: Henry Holt, 1998.
Reynolds, Ross. *Cambridge Guide to the Weather.* New York: Cambridge University Press, 2000.
Robinson, Andrew. *Earth Shock.* New York: Thames & Hudson, 1993.
Sayre, April Pulley. *El Niño and La Niña: Weather in the Headlines.* Brookfield, CT: Twenty-First Century Books, 2000.
Schneider, Stephen H., ed. *The Encyclopedia of Climate and Weather.* New York: Oxford University Press, 1996.
Simon, Seymour. *Weather.* New York: William Morrow, 1993.
Souza, Dorothy M. *Hurricanes.* Minneapolis, MN: Lerner Publications, 1994.
Stevens, William K. *The Change in the Weather: People, Weather, and the Science of Climate.* New York: Delacorte Press, 1999.
Williams, Jack. *The Weather Book.* New York: Vintage Books, 1997.

Magazine Articles

Bentley, Mace. "From 'Clippers' to 'Choo-Choos': Winter Storm Nicknames." *Weatherwise,* December 1997, p.27.

Boggs, Johnny D. "Storm of the Century." *Boys' Life,* September 2000, p.6.

Condella, Vince. "Is It Coming or Going?" *Earth,* April 1998, p.56.

"Living Safely on this Vigorous Earth: Science Refines Management of Quake and Hurricane Risks." *The Christian Science Monitor,* 31 August 2000, p.14.

Nichols, Mark. "Disturbing Forecasts: Global Warming Could Devastate Northern Habitats." *Maclean's,* 11 September 2000, p.16.

Reed. Jim. "Weather's New Outlook." *Popular Science,* August 1997, p.56.

Suplee, Curt. "El Nino/La Nina: Nature's Vicious Circle." *National Geographic*, March 1999, p.72.

Web Sites

American Meteorological Society
www.ametsoc.org/ams
This is the Web site of the largest professional meteorological society in America.

Climate Prediction Center
www.cpc.noaa.gov
This site provides long-range weather forecasts, as well as information on El Niño and La Niña, drought, and the ozone layer.

Department of Defense
www.hurricanehunters.com
Sponsored by the DOD, this site provides flight reports, weather and satellite photographs, a guide to reading weather charts, and a virtual trip into the eye of a hurricane.

Fact Sheet: Tornadoes
www.fema.gov/library/tornadof.htm
Advance planning and quick response tips for surviving a tornado, from the Fedeal Emergency Management Agency (FEMA).

National Climatic Data Center
www.ncdc.noaa.gov
This organization maintains an archive of historical weather and climate data, as well as detailed records of storm events in every state and summaries of major historical weather events.

National Weather Service
www.nws.noaa.gov
The NWS provides weather warnings, drought charts, and international climate predictions.

Storm Prediction Center
www.nssl.noaa.gov/spc
This group of National Weather Service meteorologists is responsible for the monitoring and forecasting of severe weather across the United States.

The Weather Channel
www.weather.com
This site has satellite and Doppler radar images, time lapse images, a glossary of weather terms, and current weather reports.

INDEX

NOTE: Bold page numbers indicate primary references.

A

Acid rain, **9–10**. *see also* Pollution
Adiabatic process, **10–12**
    dry adiabatic lapse rate, 12
    moist adiabatic rate, 11
Advection, **12–14**
    warm, overrunning and, 166–167
Agricultural losses
    cold wave, 47
    drought, 64–66
    frost, 88–90
    global warming, 96, 100
    hail, 211
Agriculture, cloud seeding, 251
Air, **14–17**. *see also* Atmosphere
    constituents, 239
Air circulation. *see also*
General circulation effects
    climate control, 39–40, 242
    trade winds, 225–226
Air density
    relationship to
        mirage, 150–151
        pressure and temperature,
            16–17, 196–197
Air mass, **17–19**
    classification, 18–19
        cP air, 19
        cT air, 19
        mP air, 19
        mT air, 19
    formation, 17–18
        source regions, 17
    in general, 17
Air parcel. *see also* Stability
    adiabatic process effect on, 10–12
    stability of, 196–197, 207
Air pollution. *see* Pollution
Air pressure. *see also* High pressure;
Low pressure; Wind
    factors effecting
        divergence, 60
        ocean temperature, 163–164 in
general, 16–17, 48
    relation to temperature and density,
        16–17
    hydrostatic equilibrium, 119–120
Air temperature. *see also*
Ocean temperature; Temperature
    clouds effecting, 42–43
    conditions effecting,
        16, 42–43, 99, 107, 238–239
    greenhouse effect on, 99–100
    heat index, 105–106
    regulation, 239–239, 240–241
    relationship to
        ocean temperature, 163
        pressure and density,
            16–17, 190, 196–197
        weather, 242
    volcanic activity effects, 36–37
Airplane
    hazards to
        downbursts, 64, 91–92, 211, 258
        turbulence, 145, 227–228, 256–257
    hurricane observation, 117
Airports
    anemometers, 24
    NWS forecasts for, 155
    wind shear warning systems, 64, 211
Albedo, **20–21**
Alberta clipper, **22–23**
American Meteorological Society (AMS),
contact information, 148
AMS. *see* American Meteorological Society
Anemometer, **23–24**. *see also* Wind
    types
        aerovane, 24
        pressure plate, 24
        pressure tube, 24
Antarctic. *see also* Arctic; Polar regions
    climate change, 36
    ozone hole, 169
Anticyclones. *see also* High pressure
    characteristics, 13
    monsoons and, 152–153

Apparent temperature. *see* Heat index
Arctic. *see also* Antarctic; Polar regions
    polar vortex, 170
    superior mirage, 152
Aristotle, **24–25**
Atmosphere, **26–28**. *see also* Air; Weather
    chemistry, 145
    composition, 28
    evolution of, 28
    in general, 26
    interactions with oceans, 163–164
    layers
        ionosphere, 27
        mesosphere, 27
        stratosphere, 27
        thermosphere, 27
        troposphere, 27, 239
    relationship to, hydrological cycle, 118
    scale, 144–145
        mesoscale, 145
        microscale, 145
        planetary, 145
        synoptic, 145
    sky color, 188
    stable/unstable, 14, 195–196, 196–198
    vorticity, 50–51, 182
Atmospheric models. *see also* Weather models
    weather prediction, 132, 261
Atmospheric pressure. *see* Air pressure
Aurora, **28–29**. *see also* Polar regions
    formation, 28–29
    sunspots and, 29
Australia, Brickfielder wind, 137
Automobile emissions. *see also* Cities; Pollution
    acid rain, 9–10
    pollution, 170, 171, 232

B
Ball lightning, 135. *see also* Lightning
Barometer, **30–31**. *see also* Humidity
    aneroid, 30
    mercury, 30
Bergeron-Findeisen process, 172, 178
Birds, migratory, 176
Black Forest, acid rain effects, 10
Blocking pattern, **31–32**
    effects, drought, 64–65
    omega block, 31
    Rex block, 31
Blue jets, 135
Book of Signs (Theophrastus), 25
Bora, 136
Brickfielder wind, 137
Buildings. *see also* Cities
    acid rain effect on, 10
    albedo, 21
    microclimates, 150
    urban heat island effect, 38–39, 150, 232–233
Butterfly effect, **33–34**

C
Carbon dioxide. *see also* Greenhouse gases
    atmospheric, 14, 15, 38
    relationship, greenhouse gases, 95, 96, 100
Carbon monoxide, 170
Careers, in meteorology, 147–148
Cascade mountains, rain shadow, 181
Catalina eddy, **34–36**. *see also* Wind
    causes, 34–35
    detection, 35–36
    in general, 34
CFCs. *see* Chlorofluorocarbons
Chinook wind, 136. *see also* Wind
Chlorine, 169
Chlorofluorocarbons (CFCs). *see also* Global warming; Ozone
    effects
        global warming, 96, 100
        ozone destruction, 16, 169, 170, 171
Cities. *see also* Automobile emissions; Buildings
    urban flooding, 79
    urban heat island, 38–39, 150, 232–233
Climate, **36–39**. *see also* Köppen zones; Weather
    effects on
        geographic influences, 36–37
        human influences, 38–39
        Solar radiation changes, 37–38
    in general, 36
    microclimate, 148–150
Climate change, **39–41**
    climate controls, 39–40

Index

solar energy, 39
climate zones, 40–41
Köppen zones
    continental with mild winters, 40
    continental with severe winters, 40
    dry climate, 40
    polar, 41
    tropical, 40
Climatology, 144. see also Meteorology; Weather prediction
    applications
        hurricane forecasting, 116–117
        weather prediction, 248
Cloud classification, **41–42**
    altocumulus, castellanus, 42
    banner, 43
    contrails, 43
    cumulonimbus
        formation, 43, 236–237
        fractus, 42
        lightning, 133–134
    cumulus
        humilis, 42
        lenticularis, 42
    high clouds
        cirrocumulus, 42
        cirrostratus, 42
        cirrus, 42
    low clouds
        nimbostratus, 42
        stratus, 42
    mammatus, 43
    middle clouds
        altocumulus, 42
        altostratus, 42
    noctilucent, 43
Cloud formation, **42–44**
    condensation nuclei, 44, 208
        hygroscopic, 44
    in general, 41–42, 118, 139, 197–198, 238, 239, 240, 259
    inhibition, high pressure, 110
    processes
        convection, 43
        convergence, 47
        dew point, 59
        orographic lift, 165–166
Cloud seeding, 250–251
Clouds
    electrification of, 97
    meteorological functions, 42–43
    origin, 14, 28
    radiation from, 88
    viewing, satellite photography, 82–83, 143–144
Coast. see Shoreline
Cold conveyor belt, 138
Cold front, **44–46**. see also Occluded front; Stationary front
    effects, thunderstorm trigger, 207
    formation, 77
    indicated on weather map, 44
Cold wave, **46–47**
Collision/coalescence process. see also Precipitation
    precipitation formation, 172, 174, 178
Computers. see also Weather prediction
    uses, flood prediction, 82
    weather prediction, 33–34, 146–147, 156, 159–160, 200, 215, 249, 261
        grid points, 34, 159, 261
        initial conditions, 159–160
        soundings, 196
Condensation. see also Evaporation; Moisture; Water vapor
    effects
        fog, 84–85
        hurricanes, 111
    latent heat of condensation, 153
    net, 239, 240
    process, 10–11, 43, 118, 125, 166
Conduction, 106, 107
Continental drift. see Plate tectonics
Convection
    effects, cloud formation, 43
    process, 106–107, 207
Convergence, **47–48**. see also Divergence
    divergence and, 47–48, 60
    effects
        cloud formation, 47, 199
        lake-effect snow, 130
        low pressure, 138, 140
        upper-level, 110, 128
Coriolis effect, **48–50**. see also Wind
    effect on wind, 252
    pressure gradient force and, 49
Curricula in the Atmospheric, Oceanic, Hydrologic, and Related Sciences, 148

Cyclogeneis, **50–52**
    "lee-side," 50–51
    vorticity, 50–51, 182
Cyclones. see also Extratropical cyclones; Hurricanes; Low pressure
    characteristics, 13, 31, 75
    tropical, 75
Cyclonic circulation, 77

D
Degree day, **53**
    cooling, 53
    heating, 53
Density. see Air density
Derecho, **53–55**
Desertification, **55–56**. see also Global warming
Deserts, 94
Dew, **57–58**
    formation, 57
        conditions for, 58
Dew point, 57, **58–59**. see also Frost
    defined, 43
    relationship
        dryline, 66–67
        frost, 88
Diffluence, divergence and, 60
Divergence, **59–60**. see also Convergence
    convergence and, 47–48, 60
    cyclogenesis and, 51
    diffluence and, 60
    effects, low pressure, 138, 140, 258–259
    upper-level, 128
Doppler radar, **60–63**. see also Radar; Wind profiler
    Doppler effect, 61
    in general, 60, 174, 175, 200, 253
    how it works, 61–63
Downburst, **63–64**, 91–92, 211, 258. see also Thunderstorms
    microburst, 63–64
Drought, **64–66**. see also Global warming
    causes, blocking pattern, 64–65
    measurement, Palmer Drought Severity Index, 66
    recent, Mississippi River 1988, 32
    relationship to
        desertification, 56
        heat wave, 108–109
Dry deposition, 9
Dryline, **66–67**
    locations, 66
    thunderstorms and, 67
    tracking, 66–67
Dust. see also Haze; Pollution; Volcanic activity
    condensation nuclei, 44, 101–102
    effects, visibility, 234
    particulate matter, 172
    volcanic, 36
Dust bowl, 56
Dust storm, **67–70**
    causes, 68
    effects, 68–70

E
Earth
    average temperature, 177
    magnetic field, 28–29
Earth orbit. see also Seasons
    effects
        global weather patterns, 96, 241–242
        sunlight exposure, 75, 177
    relationship to
        climate change, 37–38
        Ice age, 122
        seasons, 194
Easterly wave, **70–72**
    effects
        hurricane formation, 70–72, 112–113
        monsoon, 153
    formation, 70–72
    tracking, 70
El Niño, **72–74**. see also La Niña
    effects, 73–74, 164
    formation, 72–73
        southern oscillation and (ENSO), 73
    in general, 72
    relationship to, hurricanes, 117
Equinox, **74–75**. see also Solstice
    autumnal, 74
    solstice and, 74–75
    vernal, 74
Evaporation. see also Condensation; Water vapor

net, 239, 241
perspiration and, 105
process, 11, 43, 118, 213
Extratropical cyclones, **75–79**. see also Cyclones; Low pressure
deepening of, 77–78
nor'easters, 157–158
polar front theory, 75, 76
stages of, 76–78
variations, 78–79
Extratropical lows, 139. see also Low pressure

F
Faculae, 38
Fall wind, 136. see also Mountain winds
Fawbush, Maj. Ernest J., 223, 224
Flash flood, 79, 201, 210. see also Floods
Big Thompson River, 82
Floods, **79–84**
causes
hurricanes, 115
monsoon, 154
rain, 178
storm surge, 113–114, 158
thunderstorms, 209–210
effects, fatalities, 79, 265
flash flood, 79, 82, 201, 210
forecasting, 82–84
recent
1993 Mississippi River, 31, 80–81
1997 Red River, 81–82
Florida
lightning fatalities, 133
waterspouts, 238
Foehn winds, 136
Fog, **84–86**. see also Humidity; Water vapor
dispersal techniques, 251
effects, visibility, 234
formation, 84–85, 120–121
types
advection, 85, 86
mixing, 85–86
radiation, 85
upslope, 85
Forecast zone, 155. see also Weather prediction
Forest fire, 104, 137
Forests, acid rain effects, 10

Fossil fuels. see also Global warming; Greenhouse effect
combustion
acid rain, 9–10
air pollution, 170
carbon dioxide formation, 15
Franklin, Benjamin, **86–87**
Freeze, defined, 90
Freezing, cold wave, 46–47
Freezing rain, **87–88**. see also Sleet; Winter storms
causes, inversion, 167, 191
ice storm, 87
Friction. see also Gravity
effect on wind, 252
Front. see Cold front; Occluded front; Polar front theory; Stationary front; Warm front
Frost, 57, **88–90**. see also Dew; Dew point; Ice
relationship, dew point, 88
Frostbite, 46, **90**, 254. see also Health concerns
Fujita, Theodore, 64, **91–92**
F–Scale, 221

G
Gas law, 16–17
General circulation, **93–94**. see also Air circulation
path, 93–94
weather and, 94
Geography. see also Microclimate; Topography
effects
climate, 36–37
climate control, 39
freezing rain, 88, 191
weather, 36
Glaciers, 121
Global warming, **94–96**. see also Desertification; Drought; Fossil fuels; Greenhouse effect
causes, 15, 95–96
consequences, 94
effects, 96
in general, 94–95
Grand Forks (ND), flood, 82
Graupel, **97–99**, 133–134
riming, 98

Gravity. *see also* Friction
    effect on hydrostatic equilibrium, 119, 120
Greenhouse effect, **99–100**, 177, 241. *see also* Global warming
Greenhouse gases. *see also* Carbon dioxide; Global warming
    effects, 16, 95–96, 100
Gulf Stream, 52
    discovery, 87

H

Haboob. *see* Dust storm
Hail, **100–102**. *see also* Thunderstorms
    compared to sleet, 101, 191
    formation, 173
        thunderstorm, 211, 212
    precursors, graupel, 98, 133–134
Halo, **102–103**
    sun dogs, 103
    tangent arc, 103
Hawaii, Kona wind, 137
Haze, **103–104**. *see also* Dust; Pollution
    effects, visibility, 233–234
Health concerns. *see also* Global warming; Pollution
    flash flood, 210
    frostbite, 46, 90, 254
    heat exhaustion/heat stroke, 104–105, 106
    heat wave, 108–109
    hypothermia, 121, 254, 255
    ultraviolet radiation, 170
    wind chill, 254–255
    winter storms, 262–265
Heat. *see also* Temperature
    latent heat of condensation, 153
Heat exhaustion, **104–105**, 106. *see also* Health concerns
Heat index, **105–106**
Heat lightning, 134. *see also* Lightning
Heat stroke, **104–105**. *see also* Health concerns
Heat transfer, **106–107**. *see also* Radiation
    conduction, 106, 107
    convection, 106–107
    radiation, 106, 107
Heat wave, **108–109**
    causative factors
        soil moisture, 108–109
        topography, 108
High pressure, **109–110**. *see also* Air pressure; Anticyclones
    effects of systems, 110
    formation, convergence, 48
    life cycle of system, 109–110
    upper-level ridge, 229–230
Hinrichs, Gustavus, 54
Human influence. *see also* Fossil fuels; Global warming
    desertification, 56
    global warming, 95–96
Humidity. *see also* Barometer; Fog; Moisture; Water vapor
    relationship to
        dew point, 59
        fog, 120–121
        heat wave, 109
        hurricane formation, 112
Hurricane warning, 118. *see also* Weather prediction
Hurricanes, **110–118**. *see also* Cyclones; Low pressure
    characteristics, 139
    classification, 110
    effects
        fatalities, 79, 201
        in general, 113–115
        storm surge, 113–114, 201–201
        tornadoes, 113, 115
    eye, 110, 111
    eye wall, 110, 111
    forecasting, 116–118
        climatology, 116–117
        hurricane season, 116
    formation
        easterly wave, 70–72, 112–113
        El Niño, 73–74
        in general, 75, 112–113
        ITCZ, 113, 123–124, 125
        La Niña, 132
    in general, 110
    historical, 115–116
    seeding, 251
    structure, 110–112
Hydrocarbons, 172. *see also* Fossil fuels
Hydrological cycle, **118–119**. *see also*

Precipitation; Rain; Snow
Hydrologists, 83–84, 145
Hydrostatic equilibrium, **119–120**
    forces in
        gravity, 119, 120
        pressure gradient force, 119–120
Hygrometer, **120–121**
    types
        electrical, 102
        hair, 120
Hypothermia, **121**, 254, 255. *see also* Health concerns

I

Ice. *see also* Frost
    albedo, 21
    atmospheric record, carbon dioxide, 95
    formation, 44
Ice Age, 96, **121–123**
Ice crystal, atmospheric, 102–103
Ice crystal process. *see also* Precipitation
    precipitation formation, 172–173
Ice nuclei, 192
Ice storm, 87. *see also* Freezing rain; Sleet; Winter storms
India, monsoons, 154
Infrared radiation (IR). *see also* Radiation; Ultraviolet radiation
    in satellite photography, 83, 143–144, 146, 185
Intertropical convergence zone (ITCZ), **123–125**
    in general, 123–124
    meteorological conditions within, 124–125
    relationship to
        hurricanes, 113, 125
        trade winds, 93–94, 226–227
Inversion, effects, freezing rain, 87–88, 167, 191
IR. *see* Infrared radiation
Isopleths, **125–126**. *see also* Weather map
    names for, 126
ITCZ. *see* Intertropical convergence zone

J

Jet streak, 127
Jet stream, **127–128**. *see also* Wind
    compared to low level jet, 137
    definition, 127
    formation, 127
    in general
        jet streak, 127
        ridge, 127, 128
        trough, 127, 128, 158
    impact on weather systems, 128, 157–158
    meridional flow, 140
    patterns
        meridonal flow, 128
        zonal flow, 128
    seasons effecting, 190
    upper-level ridge, 229
    upper-level trough, 230–231
    zonal flow, 265–266

K

Köppen zones. *see also* Climate
    continental with mild winters, 40
    continental with severe winters, 40
    dry climate, 40
    polar, 41
    tropical, 40

L

La Niña, **130–132**. *see also* El Niño
    effects, 164
Lake, origin, 14
Lake Tahoe, 166
Lake-effect snow, **128–130**. *see also* Snow
    as mesoscale phenomenon, 186
    sea-effect snow, 129
    Tug Hill Plateau, 129
Land breeze, 136
Landspout, 237–238
Latent heat of condensation, 153
Lightning, **132–135**. *see also* Thunderstorms
    ball lightning, 135
    blue jets, 135
    causes, graupel, 97, 99
    effects, damage and fatalities, 132
    experiments, Benjamin Franklin, 86
    in general, 209, 211
    heat lightning, 134
    lightning bolt, 134
        dart leader, 134
        return stroke, 134
        stepped leader, 134

sprites, 135
Lightning rod, 86
Local winds, **135–137**. *see also* Mountain winds; Wind
    Brickfielder, 137
    fall wind, 136
    foehn winds, 136
    Kona, 137
    land breeze, 136
    mountain breeze, 136
    sea breeze, 135–136
    Simoom, 137
    Sundowner, 137
    valley breeze, 136
Lorenz, Edward, 33
Low level jet, **137–138**. *see also* Jet stream
    cold conveyor belt, 138
    compared to jet stream, 137
    MCCs and, 142–143
Low pressure, **138–139**. *see also* Air pressure; Cyclones; Extratropical cyclones
    deepening of, 77–78, 158
    extratropical lows, 139
    filling of, 139
    formation
        convergence/divergence, 138
        cyclogenesis, 50–52
    meteorological associations, 31
    tropical lows, 139
    upper-level trough, 230–231

M

Magnetic field, earth, 28–29
Marine layer. *see also* Shoreline
    effects, 34–35
Mauna Loa Observatory, 95
MCC. *see* Mesoscale Convective Complex
Mediterranean climate, 40
Meridional flow, **140**. *see also* Jet stream
Mesocyclone, **140–142**. *see also* Thunderstorms
    effects
        thunderstorm, 212
    tornado formation, 61–62, 142, 220–221
    formation, 141
    in general, 140–141
Mesoscale Convective Complex (MCC), 142–144. *see also* Thunderstorms
    characteristics, 142–143
    in general, 142
    tracking, 143–144
Meteorologica (Aristotle), 24, 25
Meteorologist. *see also* Weather prediction
    dynamic, 145
    tropical, 145
Meteorology, **144–148**. *see also* Climatology; Weather
    American Meteorological Society, contact information, 148
    aspects, 144–145
        climatology, 144
        paleoclimatology, 144
        weather forecasting, 144
    careers, 147–148
    in general, 144
    origin of term, 24
    tools
        human brain, 147
        radar, 145
        satellite imagery, 146
        weather observations, 146
Methane, 96, 100
Microburst, 63–64. *see also* Downburst
Microclimate, **148–150**, 186. *see also* Climate; Geography
Migratory birds, 176
Milankovitch theory, climate change, 37
Miller, Cpt. Robert C., 223, 224
Mirage, **150–152**
    superior, 152
Mississippi River
    1988 drought, 32
    1993 flood, 31, 80–81
Missouri River, flood, 81
Mistral, 136
Moisture. *see also* Condensation; Humidity; Soil moisture; Water vapor
    relationship to
        air temperature, 105
        thunderstorm formation, 207
Monsoons, **152–154**
    causes, 152–153
    desertification and, 56
    effects, 153–154
Moon, halo, 102–103
Mountain. *see also* Orographic lift

microclimates, 149
rain shadow, 180–181
Mountain winds. see also Local winds; Wind
adiabatic warming events, 12
fall wind, 136
microclimates, 149
mountain breeze, 136
turbulence, 227–228
vorticity, 50–51, 182
warming, 12, 136

N

National Oceanic and Atmospheric Administration (NOAA), 155
National Weather Service (NWS), **155–156**
advances in, 156
flood advisories, 84
in general, 155
storm advisories, 199–200
weather advisories, 155–156
weather discussions, 148
Nitrogen. see also Air; Atmosphere
atmospheric, 14, 28
Nitrogen oxides. see also Acid rain; Pollution
in acid rain, 9–10
in air pollution, 171
Nitrous oxide, 96, 100
NOAA. see National Oceanic and Atmospheric Administration
Nor'easter, **157–158**
Numerical weather prediction. see also Weather prediction
computer models and, 159–160
grid points, 34, 159, 261
initial conditions, 159–160
limits of, 160–161
NWS. see National Weather Service

O

Occluded front, **161–162**
"cold," 162
indicated on weather map, 161
"warm," 162
Ocean, **163–164**. see also Tides
airborne pollution, hygroscopic particles, 44
atmosphere-ocean interactions, 164
coastal flooding, 79–80, 82
origin, 14
relationship to weather, 163–164
Ocean currents
effects
climate control, 39
weather systems, 65, 163
influences, Coriolis effect, 50
Ocean temperature. see also Air temperature; Temperature
relationship to
air temperature, 163–164
hurricane formation, 112
La Niña, 130–131
Oklahoma City, tornado, 62
Orographic lift, **165–166**. see also Mountain
formation of clouds, rain, snow, 165–166
in general, 165
regions experiencing, 166
Overrunning, **166–167**
Oxygen. see also Air; Atmosphere
atmospheric, 14
origin, 14, 28
Ozone, **168–170**. see also Chlorofluorocarbons
destruction, 16, 169–170, 171
consequences, 170
in general, 15–16, 27, 168
as pollutant, 172
Ozone hole, 168, 169

P

Paleoclimatology, 144
Palmer Drought Severity Index, 66
Particulate matter, 172. see also Dust
Perspiration, 105, 106. see also Evaporation
PGF. see Pressure gradient force
pH scale, 9
Photodissociation, 14
Photosynthesis. see also Plants; Solar radiation
oxygen creation, 14
Plants, 14, 28, 119, 170, 233
Plasma, 28
Plate tectonics
relationship to
climate effects, 36
Ice age, 122
Polar front theory
in general, 75, 76
stages, 76–78

variations, 78–79
Polar regions. *see also* Antarctic; Arctic; Aurora
    climate, 40, 41
    sunlight exposure, 177
Polar vortex, 169–170
Pollution, **170–172**. *see also* Acid rain; Dust; Haze
    effects
        acid rain, 9–10
        visibility, 233–234
    types
        automobile emissions, 9–10, 170, 171, 232
        carbon monoxide, 170
        chloroflurocarbons, 171
        hazardous air pollutants, 171
        hydrocarbons, 172
        nitrogen oxides, 171
        ozone, 172
        particulate matter, 172
        sulfur dioxide, 171
Precipitation, **172–174**. *see also* Hydrological cycle; Rain; Snow
    detection, Doppler radar, 60
    formation, 118
        Bergeron-Findeisen process, 172, 178
        collision/coalescence process, 172, 174, 178
        ice crystal process, 172–173
    origin, 14, 28
    runoff and, 118–119
    visibility and, 234–235
Pressure. *see* Air pressure; High pressure; Low pressure
Pressure gradient force (PGF), 49

R
Radar, **174–176**. *see also* Doppler radar; Weather prediction
    how it works, 174–175
    uses
        flood prediction, 82
        meteorology, 145, 175–176, 248–249
        tracking MCCs, 144
        tracking thunderstorms, 215
Radiation, **176–177**. *see also* Solar radiation; Ultraviolet radiation
    in general, 176
    heat transfer, 106, 107
    radiation waves, 176–177
    reflected, 20, 176–177, 241
    temperature and, 177
    Van Allen radiation belts, 29
    wavelength, 179
Radiosonde, **194–195**, 214. *see also* Weather prediction
Rain, **177–179**. *see also* Freezing rain; Precipitation
    formation
        in general, 178
        orographic lift, 165–166
        urban heat island, 233
        warm advection, 167
    in general, 177–178
    in hurricane, 115
    measuring, 178
    rain-snow line, 262
Rain shadow, **180–181**
Rainbow, **179–180**
Raindrop, characteristics, 178
Rainforests, 94
Refraction, 180
Rex, Daniel F., 31
Ridge. *see also* Upper-level ridge
    movement, "retrogression," 230
Riming, 98
Rossby, Carl-Gustav, **182**
Rossby waves, 182
Runoff, 118–119. *see also* Flood

S
Sahel, desertification, 56
Santa Ana winds, 12, 136–137
Santa Barbara, Sundowner wind, 137
Satellite, **183–185**
    types
        geostationary (GOES), 183, 184–185
        polar orbiters, 183–184
Satellite photography. *see also* Weather prediction
    types
        infrared, 83, 143–144, 146, 185
        visible, 185
        water vapor imagery, 185

Index

uses
    cloud observation, 82–83, 143–144
    flood prediction, 82–83
    hurricane prediction, 117
    meteorology, 146, 248–249
    thunderstorm prediction, 214–215
    tracking MCCs, 143–144
Scales of motion, **185–187**. see also Weather systems
    mesoscale, 186
    microclimate, 186
    microscale, 186
    planetary scale, 187
    synoptic scale, 186–187
Scattering, 177, **187–188**
    Mie scattering, 188
    Rayleigh scattering, 188
Sea breeze, 135, 253–254
Sea-effect snow, 129
Seasons, **189–190**. see also Earth orbit; Equinox; Solstice
    climate control, 39, 242
Shoreline. see also Marine layer; Storm surge
    flooding, 79–80, 82
    land breeze, 136
    microclimates, 150
    sea breeze, 135–136
Silver iodide, 250–251
Simoom, 137
Skew-T diagrams, 195
Sky, blue color, 188
Sleet, **190–191**. see also Freezing rain; Winter storms
    causes, inversion, 167, 191
    compared to hail, 101, 191
    formation, 173
Snow, **192–193**. see also Lake-effect snow; Winter storms
    albedo, 20–21
    degrees of snowfall, 193
        blizzard, 193
        flurry, 193
        heavy snow, 193
        light snow, 193
        moderate snow, 193
        snow shower, 193
        snow squall, 193
        visibility and, 234–235
    depth, compared to rain, 84
    formation
        orographic lift, 165–166
        warm advection, 167
    precursors, graupel, 98
    rain-snow line, 262
    transformation to rain, 173
    wind shear and, 257
Snowflake, formation, **192–193**
Soil moisture. see also Moisture
    effects, radiation fog, 85
    relationship to
        flood, 83–84
        heat wave, 108–109
Solar radiation. see also Solstice; Sun; Ultraviolet radiation
    effects
        climate, 37–38
        climate control, 39
        greenhouse effect, 99
        wind, 252
    factors effecting
        clouds, 43
        earth orbit, 75, 177
    refraction, 180
    scattering, 177, **187–188**
    wavelength, 176, 179
Solar wind, 28–29
Solstice, **193–194**. see also Equinox; Seasons
    equinoxes and, 74–75
    in general, 190
Sounding, **194–196**. see also Weather prediction
    data use
        computer models, 196
        skew-T diagrams, 195
    in general, 194
    radiosonde, 194–195, 214
Southern oscillation event. see El Niño
SPC. see Storm Prediction Center
Sprites, 135. see also Lightning
Stability, **196–198**. see also Air parcel
    in general, 14, 195–196, 256
    jet stream and, 230
    stratosphere and, 227, 243
    thunderstorms and, 207–209, 214, 224–225
    tornadoes and, 219

Stationary front, **198–199**. see also Cold front; Occluded front; Warm front
    effects, 198–199
    formation, 198
    in general, 198
    indicated on weather map, 198
    progression, 199
Storm. see also Tropical storm; Winter storms; *specific storm types*
    formation, 190, 240
Storm chasing, **199–201**
Storm Prediction Center (SPC), 215, 224, 225
Storm spotters, 200
Storm surge, **201–202**. see also Hurricanes; Shoreline
    contributors to
        in general, 202
        hurricanes, 113–114, 201–202
        nor'easter, 158
    effects, 202
Stratosphere. see also Atmosphere
    convergence and, 48
    in general, 27, 243
Subtropical high, 226
Sulfur dioxide. see also Acid rain; Pollution
    in acid rain, 9–10
    in air pollution, 171
Summer. see also Earth orbit; Seasons
    causes, 190
    heat wave, 108–109
Sun. see also Solar radiation
    effects
        climate, 38, 241
        weather cycles, 25
    halo, 102–103
    temperature, 176
Sun dogs, 103
Sunlight. see Solar radiation
Sunspot cycle, 38
Sunspots
    auroras and, 29
    climate effects, 38
    faculae, 38
Supercooled water. see also Water
    in cloud composition, 44, 173
Synoptic scale, 145

T
Tangent arc, 103
Temperature, **203–204**. see also Air temperature; Ocean temperature
    Celsius, 204
    Fahrenheit, 203
    Kelvin, 204
    radiation and, 177
    thermometer, 205–206
    urban heat island, 38–39
Theophrastus, 25
Thermometer, **205–206**. see also Heat; Temperature
    Bi-metallic, 206
    Liquid-in-glass, 205
    placement of, 206
Thunderstorms, **206–215**. see also Lightning; Mesocyclone
    cells, 209
        supercell, 211–213, 220
    compared to MCCs, 142–143
    dust storms and, 68
    effects
        flood, 81
        flooding, 209–210
        wind shear, 258
    forecasting, 213–215
        Storm Prediction Center, 215, 224, 225
    formation
        cold fronts, 45
        derechos, 53–55
        dryline, 66, 67
        in general, 206–208
        low level jet, 137
        wind shear, 257–258
    life cycle, 208–209
    mesocyclone, 212, 220–221
    as mesoscale phenomenon, 186
    relationship to
        downburst, 63–64, 91–92, 258
        hail, 101–102
        hurricanes, 112–113, 125
        tornadoes, 62, 175, 217, 220
    stormchasing and, 199–201
    supercell, 211–213
    triggers, 207–208, 214
Tides. see also Ocean; Storm surge
    storm influence on, 158

Tinker Air Base, 223
Topography. *see also* Geography
    relationship to
        heat wave, 108
        seasonal changes, 242
Tornadoes, **216–225**
    characteristics
        "suction swaths," 91
        "suction vortices," 91
        updraft, 218
    compared to waterspouts, 237–238
    effects
        fatalities, 79, 201, 216
        in general, 221–223
    elements, 216
    forecasting, 223–225
        tornado warning, 225
        tornado watch, 225
    formation
        derechos, 54
        in general, 218–221
        hurricanes, 113, 115
        mesocyclone, 61–62, 142, 220–221
        thunderstorms,
            62, 209, 210, 212, 217
    in general, 216, 253
    measurement, F–Scale, 221–223
    occurrence, 216–217
Torricelli, Evangelista, 30
Trade winds, **225–227**. *see also* Wind
    relationship
        El Niño, 72–73
        ITCZ, 93–94
Tropical depression, 139
Tropical lows, 139. *see also* Low pressure
Tropical storm. *see also* Hurricanes; Storm
    characteristics, 139
    classification, 113
Tropical storm warning, 118
Tropical storm watch, 118
Tropical wave. *see* Easterly wave
Tropics
    climatic conditions, 40, 94, 190
    easterly waves, 71
    sunlight exposure, 177
Troposphere, 27, 239, 242–243. *see also* Atmosphere
Tug Hill Plateau, 129
Turbulence, 145, **227–228**. *see also*
Wind shear
    eddy, 227–228
    rotors, 227–228
Typhoons. *see also* Cyclones; Hurricanes
    formation, 75

U
Ultraviolet radiation (UV). *see also* Infrared radiation; Radiation; Solar radiation
    impacts
        ozone layer, 16, 169, 170, 171
        photodissociation, 14, 28
Upper-level ridge, **229–230**
    movement, "retrogression," 230
Upper-level trough, **230–231**
Urban heat island, 38–39, 150, **232–233**

V
Valley breeze, 136
Van Allen radiation belts, 29
Visibility, **233–235**
    factors effecting, 233–234
    precipitation and, 234–235
Volcanic activity. *see also* Dust
    effects
        climate change, 36
        sunset color, 188
Vorticity, 50–51, 182

W
Warm front, **235–236**. *see also* Cold front; Occluded front
    formation, 77
    indicated on weather map, 235
Water. *see also* Moisture; Supercooled water
    relationship to, hydrological cycle,
        118–119
Water vapor, **238–241**. *see also* Condensation; Evaporation; Humidity; Moisture
    atmospheric, 14, 16, 28
        condensation and evaporation, 239
        effects, 239–241
        satellite imagery, 185
    relationship
        dew point, 58
        greenhouse gases, 100
        haze, 103–104
Waterspout, **236–238**. *see also* Tornado

Weather, **241–243**. *see also* Climate; Climatology; Meteorology
    causes, 16
Weather advisories. *see* Weather prediction
Weather extremes, **243–246**
Weather forecasting. *see* Weather prediction
Weather journal, 87
Weather map, **250**
    contours, 250
    indications
        cold front, 44
        isopleths, 125–126
        occluded front, 161
        stationary front, 198
        warm front, 235
Weather models. *see also* Atmospheric models
    grid points, 34, 159, 261
Weather modification, **250–251**
    cloud seeding, 250–251
Weather observations, 146
Weather prediction, 144, **247–249**. *see also* Radar; Satellite photography
    atmospheric models, 132, 261
    climatology, 248
    computer role in, 33–34, 146–147, 156, 159–160, 200, 215
    early, 87
    forecast zone, 155
    methods, 247–248
        analogue method, 248
        persistence forecast, 248
        trend method, 247–248
    numerical, 159–161
    sounding, 194–196, 214
    thunderstorms, 213–215
    tools, 248–249
    tornadoes, 223–225
    tracking, 256
    weather advisories
        by NWS, 155–156
        severe weather advisory, 156
    winter storms, 261–262
    zone forecast, 155
Weather systems
    influences
        Coriolis effect, 48–49
        oceans, 163–164
    movement, 110, 182

    causes, 65
    scales of motion, 185–187
    tracking, 256
Wildlife, acid rain effect on, 9–10
Wind, **252–254**. *see also* Jet stream; Local winds; Mountain winds; Trade winds
    adiabatic warming events, 12
    Catalina eddy, 34–36
    Coriolis effect on, 48–50, 252
    derechos, 53–55
    formation, 16, 190, 252
    hurricane, 114–115
    measurement, 252–253
    patterns, 253–254
    Santa Ana, 12, 136–137
Wind chill, **254–255**. *see also* Health concerns
Wind profiler, **255–256**. *see also* Doppler radar
Wind shear, **256–258**. *see also* Turbulence
    relationship to
        mesocyclone, 141
        tornado formation, 225
        turbulence, 228
Wind shear warning systems, 64, 92
Wind speed
    anemometers, 23–24
    measurement, 252–253

Winter
    causes, 189
    cold wave, 46–47
Winter storms, **258–265**. *see also* Storm
    formation, 258–261
    ice storm, 87
    prediction, 261–262
    rain-snow line, 262
    Superstorm, 262–265

Z
Zonal flow, **265–266**. *see also* Jet stream

About the Author

Paul Stein has a B.S. in Meteorology from Pennsylvania State University. He has eight years of experience as a weather forecaster, most recently as a Senior Meteorologist for The Weather Channel. Currently, he develops computerized weather forecasting processes and products.

Photo Credits

p. 9 © Telegraph Colour Library/FPG; p. 11 © David McGlynn/FPG; p. 15 © CORBIS; p. 16, 43, 54, 57, 97, 136, 164, 179, 234, 240, 243, 253, 257, 263 © Superstock; p. 20 © Jim Steinberg/Photo Researchers, Inc.; p. 23, 61, 111, 184, 200, 211, 213 © Weatherstock; p. 25 © FPG International; p. 30 David Lees/CORBIS; p. 36 © Gary Braasch/CORBIS; p. 37, 122, 146 © PhotoDisc; p. 38 © Ray Pfortner/Peter Arnold, Inc.; p. 41 © Magrath/Folsom/ Science Photo Library/Photo Researchers, Inc.; p. 51, 217, 218, 222, 238 © Warren Faidley/ International Stock; p. 52 © Arvil A. Daniels/ScienceSource/Photo Researchers, Inc.; p. 62 © Everett C. Johnson/Uniphoto; p. 65 © Scott Camazine/Photo Researchers, Inc.; p. 69 © Jim Richardson/CORBIS; p. 73 © Philip Wallick/International Stock; p. 74 © Mark E. Gibson/ International Stock; p. 77, 116, 126, 147, 175, 214, 229 © NOAA Photo Library/NOAA Central Library; p. 79, 210 © Wayne Aldridge/International Stock; p. 80 © Bob Firth/ International Stock; p. 82 260 © Llewellyn/Uniphoto; p. 85 © Russell D. Curtis/Photo Researchers, Inc.; p. 86 © Archive Photos; p. 89 © George Lepp/CORBIS; p. 91 © VCG/ FPG; p. 95 © Frank Lane Picture Agency/CORBIS; p. 98 © Nuridsany et Pérennou/Photo Researchers, Inc.; p. 101 © Gregory G. Dimijian/Photo Researchers, Inc.; p. 102 © Corps Photo Collection /NOAA Photo Library/NOAA Central Library; p. 103 © Will & Deni McIntyre/ Photo Researchers, Inc.; p. 107 © Robert Arakaki/International Stock; p. 114 © Uniphoto; p. 118 © Wayne Lawler/CORBIS; p. 121 © Christiana Carvalho; Frank Lane Picture Agency/CORBIS; p. 124, 133, 197 © 1999 Artville LLC; p. 131 © Renee Purse/Photo Researchers, Inc.; p. 139 © Nancy Setton/Photo Researchers, Inc.; p. 141 © Faidley/Agliolo/International Stock; p. 144 © Tom Murphy/Superstock; p. 145 © Christopher Harris/Uniphoto; p. 149 © David Frazier/The Image Works; p. 150 © Richard Hamilton Smith/CORBIS; p. 154 © Bruno Barbey/Magnum/PictureQuest; p. 160 © Image Courtesy of Center for Ocean-Land-Atmosphere Studies; p. 168, 183, 247 © Space Photo Collection/NOAA Photo Library/ NOAA Central Library; p. 171 © G. Büttner/ Naturbild/OKAPIA/Photo Researchers, Inc.; p. 173 © John Wilkes/FPG; p. 186 © Franke Keating/Photo Researchers, Inc.; p. 187 © Telegraph Colour/FPG; p. 192 © Robert Pickett/CORBIS; p. 195 © National Weather Service, Buffalo, New York; p. 203 © Bettmann/CORBIS; p. 205 © Frank Siteman/ Uniphoto; p. 208 © Keith Kent/Science Photo Library/Photo Researchers, Inc.; p. 220 © Howard Bluestein/Photo Researchers, Inc.; p. 224, 228 © NSSL Photo Collection/NOAA Photo Library/NOAA Central Library; p. 232 © Margaret McCarthy/Peter Arnold, Inc.; p. 237 © J.H. Golden/Photo Researchers, Inc.; p. 249 © Mark and Audrey Gibson/PictureQuest; p. 251 © Mark Sephenson/CORBIS; p. 259 © Jim Grace/Photo Researchers, Inc.; p. 264 © Peter Beck/Uniphoto.

LIBRARY
Burnsview Sec. School
7658 - 112th Street
DELTA, B.C.
V4C 4V8